1

"Behind it all,
  Is surely an idea so simple,
  So beautiful,
  So compelling,
  That when we grasp it,
  We will all say to each other,
  How could it have been otherwise?"

John Archibald Wheeler

DEDICATED TO MY MOTHER CAROL AND MY UNCLE ROY

Support from the following friends is gratefully acknowledged

Robert and Gloria Potter

James J Brennan

Robert F Bennion

And Astrid Lindholm, my dear wife who each day must await my return to reality after mentally absenting myself to ponder some new twist on an old cosmic mystery

Preface to 2nd Edition

The riddle of the universe is integrally tied to space and time. The peculiarities of these abstractions determine both the mannerisms of matter, and the influence of physically separated objects upon one another. That the insightful theories advanced by Newton and Einstein still leave unanswered questions, is the motivation for this treatise.

In contrast to scientific journals that report day-to-day progress and failures in the efforts to resolve scientific mysteries, this is a story in hindsight, a post retirement reflection upon what the scientific community has overlooked as an explanation of Gravity and Inertia. For the author, the quest began on a cold winter's night in the mountains of Utah. The family vacation home had become a favorite place to explore new ideas with old friends. The fire had died low, the hour was late, colloquy had drifted to the enigma of "*gravity*." Conjectures provoked comments and more conjectures; one in particular, became a quest.

The prevailing theory of gravity is based upon curvature, but the amount of curvature depends upon the measured value of **G**. The predictive power of General Relativity rests entirely upon the same mysterious acceleration factor as Newtonian theory. To this extent, both theories are incomplete. As once lamented by Stephen Hawking: "…..we have two theories of gravity, but neither can predict its strength." What better way to spend a retirement, tracking down the source of **G**? No schedules to meet, no preceptors to please, working at leisure, free to make mistakes in private. Indeed, the mistakes have been many, starting with a strong faith in the pre 1998 "Standard Model. Following the interpretation of the supernova studies, the universe was promoted from deceleration to acceleration. And the mystery of the **G** field, simply a hitherto overlooked apotheosis of accelerating expansion.

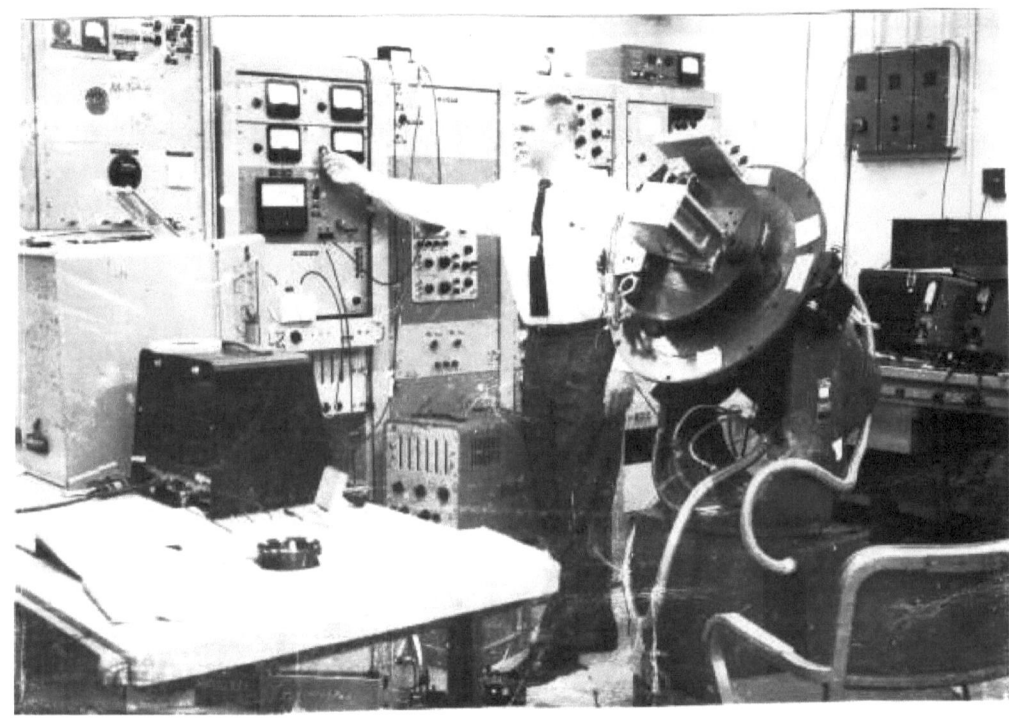

<u>Inertial Guidance Laboratory, TRW Space Technology Systems.</u> Circa 1965 photo of the author testing his fail-safe design for moderating descent velocity during Lunar landings. *"The gratification in having contributed to the success of the Apollo program is reward for a lifetime."*

Bruce D. Jimerson

Academic Background:

BSEE - UCLA, 1959
MSEE - UCLA 1963,
JURIS DOCTORATE - Loyola 1967.

FUNDAMENTAL FORCES FROM FIRST PRINCIPLES

Introduction

Nature's laws reveal an intrinsic communion between space and time. The ratio of one to the other is given the symbol 'c' commonly consociated with the speed of light. In reality, 'c' is panoptic. Energy and mass are linked by c^2 and, as developed so also are gravitational and electric fields.[1]

The conditions for adiabatic self-creation are simple but sever. Momentum and energy must be zero within the limits of the uncertainty principle. Zero carries with it the implication of the whole in perpetual communication with its parts. What then is the universal commutator, that which underwrites the laws of physics? In the words of Richard Feynman: "....*it is the space, the framework into which the physics is put .*"

Modern physics does not consider empty space as a medium. In lieu of force and motion, static curvature and stress are taken as descriptors. But without a substantive physiology, these abstractions have no meaning, an essence stressed by Einstein in his later years. If space is conceded to be nothing, can it still have dynamic functionality? Posing the question, is to suggest an analogical response. As a vacuum powered bellows motor is driven by external atmospheric pressure, so also will a void in tension beget matter and motion.

Einstein realized that masses do not act directly upon one another. Some sort of medium was needed. To fill the void (so to speak), he renamed the ether "space," and for the lack of a field, postulated it would be conditioned by inert matter. The theory did not predict the extent of the distortion nor did it explain how inert mass acts upon static space to make it "unusual." Moreover, to tell space how and where to curve requires an underlying informational field. Only after the measured value of the acceleration coefficient **G** (also known as the gravitational constant) is inserted into the equations, does the theory make testable predictions. The most important aspect of gravity, the cause of the global **G** field is not addressed in General Relativity.

Symmetry of action follows from the proposition that momentum is conserved when the roles of space and mass are reversed. Just as 2nd law Newtonian forces are proportional to their inertial content, so also must spatial acceleration be opposed by spatial reaction. Local 'g' fields are proportional to the masses with which they are associated, electric fields are proportional to the charge quantum 'q.' Both forces are coextensive with the extent of the universe and as to be shown, they are the devises thereof. Emergent fields created by expansion, however, does not imply physical motion of mass-less space. What is observed, as in General Relativity, is that the coordinate values specifying the reactive force of a test mass is altered by the introduction of a nearby mass in relation to the expanding coordinate system.

---

[1] "Field Forces from First Principles" The Common Cause of Natures Two Long Range Fields, B. Jimerson, 2014, Amazon Books.

FORWARD

"In questions of science, the authority of a thousand is not worth the Humble reasoning of a single individual."

Galileo Galilei

The universe is constructed along simple lines, comprehensible to some extent by human minds. Natures recurring patterns are forged from principles applicable at all times and in all places. From this spatiotemporal consistency, theories emerged to explain matter, inertia, action and reaction. Those that survived experimental scrutiny became the laws of mechanical physics. Once embraced by the stewards of a particular discipline, they were not to be easily undone. Yet as history shows, scientific reforms do occur, often being brought about by an *"upstart crow"* from another field with the temerity to interrogate treasured doctrines. And so it is with our cast of characters, and the parts they played in reforming scientific thought.

Aristarchus of Samos appears to be the first of the Ancient Greeks to expound the schema of a sun-centered universe. As with other credo's denigrative of terrestrial importance, it would be later condemned as heretical by Christian authorities who embraced Ptolomies' *"geocentric universe"* as providential. It was only on his deathbed in 1543, that Polish astronomer, Nicolaus Copernicus, dared propose a rival theory that elegantly de-mystified retrograde planetary motion. After decades of repression, compelling evidence came with the invention of the telescope.

In 1610, Galileo Galilei, turns his newly invented spyglass toward the night sky and observes the moons of Jupiter. His courageous pronouncement that: "Everything does not revolve about the earth" earns him a trip to the Inquisition. Galileo's most important emendation to physics, however, received less reproach. By showing that gravity accelerates all weights equally, he disproved Aristotle's *"tenants of motion"* and laid the cornerstone for the laws of classical mechanics.

Born posthumously on Christmas day in the year that Galileo died, the only child of an illiterate yeoman in Woolsthorpe England, Isaac Newton would survive premature birth and physical frailty to become the mental giant who discovered the relationship between accelerating motion and inertial reaction. He also identified the local impetus which called "apples-to-fall" with the force that coerced planets to follow elliptical paths around the Sun. Newton's "Law of Inertia and his "Law of Gravity," were regarded as separate and distinct forces for 25 decades. But a challenge to that proposition would arise in the early years of the 20th century with the publication of several remarkable papers authored by an obscure clerk working in the Swiss Patent Office.

Albert Einstein, once thought by his parents to be retarded, ignited a scientific reformation based upon the counterintuitive proposition that light speed is the same for all observers. The novel *"Theory of Special Relativity"* offered a new exposition of space and time. The long sought existence of a luminiferous ether was rendered moot, temporal intervals subjective and simultaneity frame dependent. The notion of a physical reality based upon common sense awaited even further dismantling by what was soon to follow.

In 1907 Einstein used the Special Theory to formulate the $\mathbf{E = mc^2}$ relationship between energy and mass. During the same year, his former professor, Herman Minkowski, was at work amalgamating temporal and spatial intervals into a single conceptual entity. Both unifications would have significant impact upon how the universe would be interpreted, but neither provided insight into Galileo's discovery that all weights fell at the same rate. That nexus would be later answered by Einstein in a single inspirational moment, with the realization that gravitational and inertial mass, are one-in-the-same. Known thereafter as the *"Equivalence Principle,"* the idea of a separate gravitational force to direct the motion of objects in a gravitational field was replaced by mass induced spatial curvature. With the publication of General Relativity in 1916-17, the paths of falling apples and orbiting planets were reduced to a single equation.

Bending space was a new job for inertial mass, its sole previous duty being that of opposing velocity change. To explain the affinity between masses, Einstein adopted a positively curved Riemannian manifold to describe static space. To prevent gravitational collapse, a cosmological factor (symbolized by the Greek letter Λ) was introduced in the final draft. Shortly thereafter, Dutch astronomer, Willem de Sitter, noticed that General Relativity admitted an expanding solution based solely upon Λ. The concept of 'negative pressure' being yet to be imagined, de Sitter's universe was devoid of both mass and pressure. The idea of an exponentially dilating vacuum would likely have been only a curiosity save for the arresting fact that astronomers had begun to observe unexplained red shifts in the spectrums of faint galaxies.

Comes then Alexander Friedmann, a Russian mathematician with a density dependent expansion theory, also consistent with General Relativity. Although Friedmann's 1922 publication went largely unnoticed at the time, the same equations would be later re-discovered by Belgian Priest, George Lemaitre. In the mind of many, expansion carried the implication of a beginning, resulting in the Pope's pronouncement that science had proved genesis. While the ontology fell far short of validating biblical doctrine, de Sitter, Friedman and Lemaitre, did establish expansion as a plausibility – a reality later confirmed by an America who would forsake his legal career to study the stars.

Edwin Hubble, the lawyer turned astronomer, is appropriately hailed as the man who measured the universe. Using the 100-inch Hooker Reflector on Mount Wilson, Hubble and his assistant, Milton Humason, collected the galactic red-shift and luminosity data that led to the "Theory of Expansion."

Surprisingly, the discovery did not adversely impact the geometric construct upon which the General Theory is founded. Einstein, however, was remorse. In having pursued a course he himself had wrongfully set, the opportunity to predict expansion prior to its experimental detection, had been lost. With no apparent need for a cosmological constant Λ, he suggested it be "awayed"

Spatial expansion, however, presented a balancing puzzle of its own. Dilation rate and gravitational retardation due to matter density, appeared miraculously equalized. Could the disparaged Λ find employment in a dynamic universe? To cancel gravity, Einstein's Λ needed to have a value of $3H^2$ where $H = 1/\tau$ (τ is the Hubble time about 13.8 billion years). In most genesis cosmologies, H is a slowly changing variable that determines the scale (R) and the time span (τ).

Enter next, Richard Feynman, one of many mathematical physicist's attempting to unify classical gravity with quantum theory. In a series of lectures given at Caltech in 1962, he posed the question thus:

1) Gravitation is a new field of its own, unlike anything else, or

2) Gravity is a consequence of something already known but incorrectly perceived.

The empirical support for a theory of gravitons congruent with the successful predictions of General Relativity, was then, and is today, still missing. Feynman's lecture series led to questions but not answers. Does a spin-2 gravity particle really exist, and if not, how is it that space opposes acceleration in one experiment and holds masses together in another. Einstein had settled upon curvature, in his words, the proximity of matter made the vacuum unusual. But matter altering spatial geometry, requires new physics. For Richard Feynman, the idea that gravity may be something already known but incorrectly perceived, seemed to always be with him. It is in his musings rather than any contribution to quantum gravity theory, that Feynman is to be credited with a revealing comment:

"One very important feature of pseudo forces is that they are always proportional to the masses. The same is true of gravity. The possibility exists therefore that gravity itself is a pseudo force. Is it not possible that perhaps gravitation is due simply to the fact we do not have the right coordinate system?"

Richard Feynman[2]

---

[2] *Lectures on Physics, Vol I at page 12-11*

Feynman's Force -- Folly or Foresight

Richard Feynman referred to inertial reactions as *"pseudo forces."* He went on to illustrate how acceleration(s) could be related to spatial curvature. But if gravity is a pseudo force arising from acceleration, nothing more is needed to explain local **'g'** fields as consequent. The path from Newton to Einstein, to Friedman to Feynman, ironically returns back to Newton *a la* his 2nd Law. If inertial reaction provokes a change in the local characteristics of expanding space, how might it be perceived?

In 1916, the dynamic expansion field remained yet to be discovered. To fill the void (so to speak), Einstein invented static curvature. Curiously, Newton's 2nd law offered a ready-made dynamic solution. That it is silent on symmetry renders the reactionary force no less mysterious if accelerating motion is assigned to expanding space rather than a change in the velocity of mass. To follow the path suggested by Feynman through to the derivation of **'g'** forces as inertial reactions, it will also be necessary to borrow from Einstein along the way.

While working out General Relativity, Einstein pondered what properties the universe must have to prevent detection of absolute motion. What followed was the *"Principle of Relative Acceleration."* The force felt by an accelerating rocket ship would be no different if the rocket were at rest in an accelerated universe. Would the same be true for isotropic coordinate expansion?

Hubble's data revealing the correspondence between albido and red shift was interpreted by Howard Robertson to be the result of stretched space. From this, Robertson (not Hubble) deduced the velocity-distance law, **[v = Hr]**, now considered the central law of modern cosmology, and one of the most important discoveries of the 20th century. At any moment of cosmic time, the recessional velocity of space increases linearly with distance, a fact having determinative significance for gravity in that volumetric expansion increases geometrically. Uniform spherical masses immersed in a uniform radial divergence flux should experience isotropic counter forces. Reversing the roles of space and mass unifies gravity with inertia without invoking spatial curvature. Taking local **'g'** fields as reactionary pseudo forces, and global coordinate expansion as provocateur, gravity reduces to **[F = ma]**.

To complete the formulization of gravity as a pseudo force within Feynman's denotation, the acceleration field must be quantified. Help comes from two British cosmologists. In 1934, William McCrea and Edward Milne, applied basic Newtonian energy relationships to re-create the gravitational equations previously extracted from General Relativity:

$$\ddot{\mathbf{R}} = -\frac{4\pi\mathbf{G}}{3}\left[\rho_u + \frac{3\mathbf{P}_s}{c^2}\right]\mathbf{R} + \frac{\Lambda\mathbf{R}}{3} \tag{1}$$

$$\left[\frac{dR}{dt}\right]^2 = \frac{8\pi\mathbf{G}\mathbf{R}^2\rho_u}{3} + \frac{\Lambda\mathbf{R}^2}{3} - \frac{kc^2}{R^2} \tag{2}$$

9

Where **R** is the Hubble radius, **k** is the curvature constant, **G** is Newton's gravitational constant, Λ is Einstein's cosmological constant, and the double superior point over **R** indicates radial acceleration. That the same gravitational equations can be derived from energy relationships raises the question of whether mass induced curvature can be sustained as the best way to describe gravity. Both Milne and McCrea expressed doubts.

McCrea later devised a cosmological mass creating algorithm independent of General Relativity.[3] Starting with the proposition that an elastic material produces heat when stretched, McCrea reasoned that an empty volume would create positive energy if it were further expanded. For a universe in tension, the theory is a straightforward application of the first law of thermodynamics. Added energy has mass, during expansion the energy released could take form as newly created particles or as enhanced inertia. In the 1950's, this notion was seized upon by Fred Hoyle and other supporters of *"Steady State Theory"* as well as later Theorists such as Alan Guth who adapted it to explain the Hubble mass as an *Inflationary Stage* of rapid initial growth. While Steady State theory was largely discredited by the discovery of Cosmic Background Radiation (CBR), expansion created energy remains a part of modern cosmology. Herein, it will prove invaluable to particle formation and inertial enhancement.

While the theory of a universe *ex nihilo* is not new, herein, the idea is taken as axiomatic. For the Hubble sphere, Newton's 2nd law can then be expressed in terms of M_u as denoting all forms of energy contained thereby:

$$M_u(dv/dt) = v(dM_u/dt) \qquad (3)$$

Because matter is composed of particles held together by electrical and quantum forces at small scales and by gravitational attraction on the large scale, spatial expansion does not cause particles to separate except at distances where the $1/r^2$ attractive force of gravitation is less than the cosmological expansion force. Spatial divergence, however, is functionally operative at even the smallest distances. The manifest of expansion created acceleration acting upon non-expandable matter is the gravitational gradient.

Both Newton and Einstein, believed gravitational attraction depended from some form of continuous action. While General Relativity offers a geometric ontology that explains motion of masses, it leaves unanswered the question as to what determines **G**. As Stephen Hawking once lamented, *"we have two theories of gravity, but neither can predict its strength."*

---

[3] What is consternating is that McCrea and Milne could synthesize Einstein's theory of gravity from Classical Dynamics." While Newton's definition of force as "rate of change of momentum" is as valid today as ever, one would expect radiation and relativistic mass to appear explicitly somewhere in the formulation. But alas, equations (1) and (2) are no different than those derived from General Relativity. This suggests, that for gravitational purposes, all forms of energy can be expressed in terms of their equivalent E/c^2 mass and amalgamated into the definition of the density term ρ_u. The underlying principle of curvature as well as time dilation, relativistic mass, and expansion itself, are then relatable by energy interdependence.

Mass, gravity, space and time are intimately entwined. As one of nature's two long range forces, gravity is coextensive with space and by reasonable account, should in some way depend from Hubble parameters. Introduce next, the Nineteenth Century physicist, Ernst Mach. Although not the first to propose inertia to be the result of other mass, he has been given that distinction. Einstein, in initially attempting to incorporate the concept as a rudiment Relativity, referred to the dependence of inertia upon other matter in the universe as "Mach's Principle." Although later changing his mind [likely because instantaneous communication appeared to be required], the association persisted. Let us examine the Theory in terms of what we think we know.

Both Mach and Einstein rejected inertial reactance as intrinsic to the mass being accelerated. But to admit a cosmological source within the predicate of Mach's Principle, we must take the path not taken by Einstein. Specifically the issue of how space instantaneously communicates other matter must be addressed.

Figure 1 depicts a 2 step transformation from fully homogenized 3-D Hubble sphere (yellow) to a flat plane (gray). Provisional specifications for Hubble mass and size are hand picked for compliance with the objective of determining the dynamic properties of space by the methods studied herein. That these values do not offend empirically determined parameters is an obvious criteria, *a fortiori*, the argument they are the only values satisfying known criteria is later promoted. Taking Hubble radius R_3 as **1.3 x 10^{26} meters** and Hubble mass M_u as **1.45 x 10^{53} kgm** (collectively all forms of mass-energy contributed by the total number of particles as if each were sufficiently separated from the others to have negligible gravitational interaction).

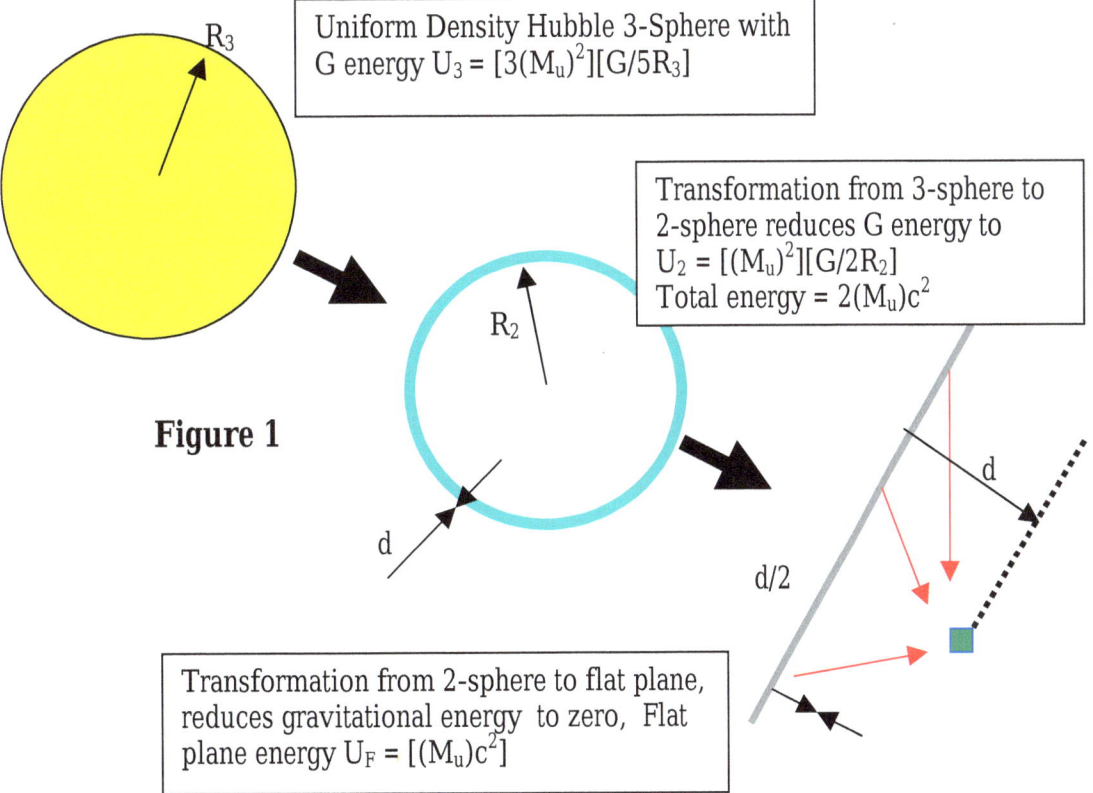

R_3

Uniform Density Hubble 3-Sphere with
G energy $U_3 = [3(M_u)^2][G/5R_3]$

Transformation from 3-sphere to
2-sphere reduces G energy to
$U_2 = [(M_u)^2][G/2R_2]$
Total energy = $2(M_u)c^2$

R_2

Figure 1

d

d

d/2

Transformation from 2-sphere to flat plane,
reduces gravitational energy to zero, Flat
plane energy $U_F = [(M_u)c^2]$

To transform a 3-sphere Hubble universe to an equivalent 2-sphere, all forms of energy contained within the 3-sphere must be included. Specifically, conversion of the Hubble to a gravitationally equivalent 2-sphere, must account for the decrease in the energy deficit from $[U_3 = 3M_u^2G/5R_3]$ to $[U_2 = M_u^2G/2R_2]$ in formulating an expression for the 2-sphere gravitational energy. Since M_u and G are premised as unchanged by transformation (the object being to determine what changes are required in the 2-sphere structure to maintain the gravitational affect of the M_uG product unchanged after transformation). A 2-sphere of the same radius ($R_2 = R_3$) will have a gravitational energy deficit of (5/6). To maintain net force at distance 'd' unchanged, U_2 must equal U_3:

$$M_u^2G/2R_2 = 3M_u^2G/5R_3$$

$$R_2 = (5/6)R_3 = 1.08 \times 10^{26} \text{ meters} \qquad (4)$$

In order to equalize 2-sphere gravitational energy with 3-sphere gravitational energy, the 2-sphere radius R_2 must be reduced by (5/6). [The attractive force created by a 3-sphere at distance $d > R_3$ from the mass center will be the same as that of a 2-sphere at the same distance 'd' if $R_2 = (5/6)R_3$]. Gravitational energy is negative, thence for a zero energy universe, 2-sphere energy M_uc^2 must be equal to the 3-sphere M_uc^2 from which it was transformed

For a 2-sphere, net gravitational force acting upon a single interior mass is zero, whereas the force on the outside of the shell is the same as if the entire mass-energy were concentrated at the center of the sphere. Thus, for an external mass M_x proximate to the shell, the gravitational force will be $4\pi G\sigma_2(M_x)$ where $\sigma_2 = 2M_uc^2/4\pi R_2^2$. The 2nd phase of transformation shown in **Figure 1**, fillets the 2-sphere density into a flat plane (shown edge-on a gray line). Mathematically this proceeds by letting R_2 approach infinity so one might expect the force on M_e near the surface of the flat plane to be the same as if the mass is distributed as a spherical shell. The force, however, is reduced by half (a fact which has led to some consternation if account is not taken of the gravitational energy deficit when transforming 2-spheres to flat planes). The gravitational energy $M_u^2G/2R_2$ is lost during transformation from 2-sphere to flat plane. The new force acting upon M_e is formulated as $4\pi G\sigma_u(M_e)$ where the energy density σ_u is ½(σ_2). Density σ_u for an infinite flat plane therefore reduces to non-gravitational energy divided by area, $(M_u/4\pi R_2^2)$. Being a global scalar σ_U will have the same value throughout the universe and exert the same influence at all places[4]

The purpose of the double transformation from 3-sphere to flat plane approaching infinite extent now becomes apparent. The force of an infinite flat plane is the same at any distance 'd' from the plane. G can now be expressed in terms of a ubiquitous scalar density function σ_u as:

$$G = F/M_e(4\pi\sigma_u) \qquad (5)$$

---

[4] Appendix A-1 details the 2-sphere to plane transformation per Feynman.

The gravitational force (red arrows) acting upon mass M_e (green) at any distance 'd' from the double transformed sphere to plane (gray) can be calculated straightaway by treating the universe as an ever-present sigma energy density surface $\sigma_u = M_u/4\pi(R_2)^2$. From (5), the opportunity to express G in terms of Hubble parameters follows directly from Newton's second law. To complete the formulization in terms of isotropic global acceleration, and to put all this in place for what follows, some observations are in order.

Average density of matter is small, yet the universe strenuously opposes acceleration. That the force of an infinite plane is independent of distance, permits the plane to be considered instantiated opposition to accelerating mass.

Inertial reactance *a la* Mach depends upon the totality of cosmic mass, but by double transformation of the entire mass M_u to a plane, we have created a mathematical equivalent which expresses volumetric density as a uniform scalar density reactance field. Density expressed as σ_u rather than p_u properly reflects the action of Machian mass as a fractional participant in creating local inertial reaction. The actual contribution will depend upon the ratio of local mass to cosmic mass, the latter herein assumed to be in the range of ($4\pi R_2^2$ **kgm/m**^{2}). The fractional share is based upon the area over which the force is distributed. This is the consequence of dividing total mass by the area over which it is distributed [To scale cosmic mass M_u as a constant reactionary at all locations, it is divided by the area $4\pi R_2^2$]. Newtonian reactionary force will be instantaneous because the reactionary scalar wall σ_u is anywhere and everywhere. That the force of an infinite plane does not depend upon the distance will be understood from the Appendix I, in words, the further away the mass Me from a particular plane, the greater the number of force vectors pointing in the direction of the plane. Indeed, in lieu of a single plane, σ_u can be a composition of many low density planes so long as the operative density is unchanged. Since each infinite plane adds an equal amount to the force independent of distance, the number of planes can be coextensive with the Hubble scale. The universe from a gravitational perspective thus models as an enumerable number of infinitesimally low density sheets that collectively bring about an operative scalar density σ_u.

Double transformation from 3-sphere → 2-sphere → N-plane, conflates Hubble energy content ($\rho_u V$) into a mathematically equivalent scalar density operative σ_u. The instantaneous dependence of inertial reaction upon the totality of mass scattered throughout the cosmos can now be understood without offending Special Relativity. Mach's opus leads to inertial values consistent with Newton's 2nd law. Simply stated, the inertial property of an individual mass M_x depends in part upon all other matter scattered throughout the universe, i.e., $M_x \propto M_u$,

$$M_x = K(M_u) = (n/A_u)M_u = = n(\sigma_u) \tag{6}$$

where 'K' is the proportionality constant between M_x and M_u and 'n' is the number of square meters M_x represents in **kgm** in relation to the entire area of the plane ($4\pi R_2^2$). By example, the earths mass and Hubble scale estimates proposed for **Figure 1**, the earth's 'n' factor area is **5.98 x 10^{24} m^2** on the σ_u plane.

From an adaptation of a theorem due to Gauss, the flux produced by expansion of a uniform spherical volume of density ρ is equivalent to the normal component of the flux summed over the surface density σ (For the Hubble, ρ and σ are subscripted "**u**")

$$\iiint_V \rho \,(\mathbf{dV}) = \iint_S \sigma(\mathbf{dA}) \tag{7}$$

In words, Gauss's divergence theorem relates the integral over the volume of the surface that contains the divergence to the flux exiting across the surface that contains the volume.[5] Conceptually, the notion of a Gaussian surface as a field measuring device will be extensively implemented throughout this thesis. The relationship between surface density σ and volumetric density ρ for a uniform spherical distribution of matter follows directly from (7) wherein the integral over the differential volume $\mathbf{dV = (4/3)\pi r^3}$ and the integral over the Gaussian surface element $\mathbf{dA = 4\pi r^2}$, hence:

$$\sigma = \rho(r/3) \tag{8}$$

where the '**r**' value in (8) will represent the 2-sphere value be addressed when making transformations involving gravitational deficits.

To apply Gauss's Theorem to an expanding Hubble sphere, summation is taken over the exiting recessional spatial flux with the object of determining flux per unit area at the instant of coincidence with a fixed Gaussian surround of radius $\mathbf{R_S}$.

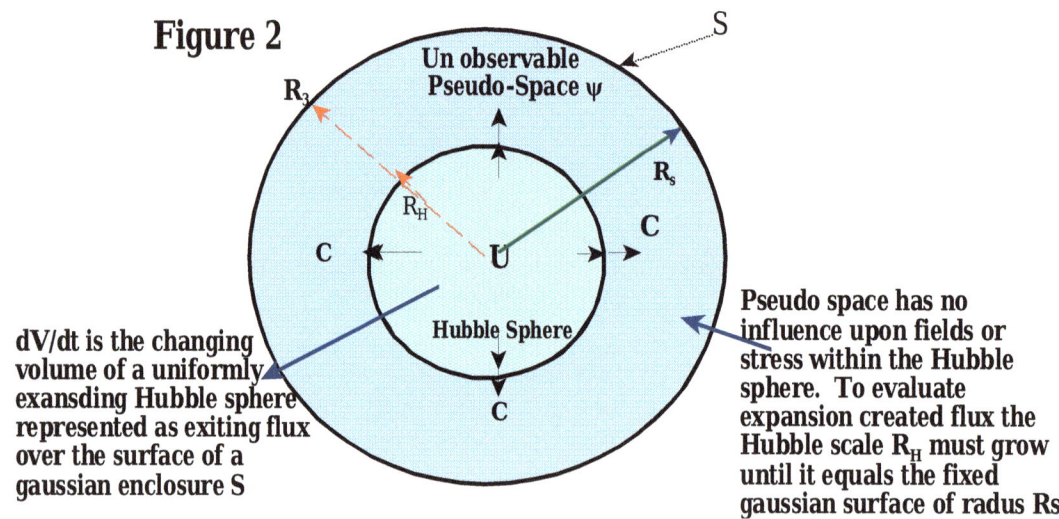

Figure 2

Un observable Pseudo-Space ψ

R_3

R_s

R_H

C   C

U

Hubble Sphere

C

S

dV/dt is the changing volume of a uniformly exansding Hubble sphere represented as exiting flux over the surface of a gaussian enclosure S

Pseudo space has no influence upon fields or stress within the Hubble sphere. To evaluate expansion created flux the Hubble scale R_H must grow until it equals the fixed gaussian surface of radius Rs

---

[5] First elaborated by the 18th Century mathematician Carl Friedrich Gauss, sometimes referred to as *Princeps Mathematicorum* (The Prince of mathematics).

14

In **Figure 2**, the radius of the 3-D Hubble sphere is subscripted R_3 and the fixed Gaussian sphere R_S. When the Hubble volume grows to equal that of the Gaussian Surround 'S' then for uniformly expanding space, the rate of change of spatial volume $dV/dt = 4\pi(R_3)^2 c$ as indicated by the arrow dV/dt denoting spatial volume per second flowing across the Gaussian surround. While the actual size of the Hubble scale R_3 as an observational limit has no physical significance in determining gravity, the value of R_3 is a convenient yardstick where, by definition, the recessional velocity is 'c.'

When **S** is coincident with R_3, the flux exiting across **S** is a measure of the rate at which space is expanding. Spatial exit flux includes that which is added to the volumetric growth of the Hubble sphere (which grows at velocity 'c') plus or minus any gain or loss due to the recessional velocity of space relative to the Hubble velocity 'c.' In a slowing universe, spatial divergence will be subluminal, in an accelerating universe, spatial flux exiting across 'S' will be transluminal. Thus, if **V** is the volume of a Hubble space expanding at uniform velocity 'c' then (9.2) is apropos and volumetric acceleration is given by the first term of (9.3). When radial rate of growth is increasing [$\ddot{R}$ positive] then (9.3) includes an additional term.

$$V = \frac{4}{3}\pi R^3 \ldots\ldots\ldots\ldots\ldots\ldots\ldots\ldots\ldots\ldots\ldots\ldots\ldots\ldots\ldots\ldots\ldots(9.1)$$

$$\dot{V} = (4\pi R^2)\dot{R}\ldots\ldots\ldots\ldots\ldots\ldots\ldots\ldots\ldots\ldots\ldots\ldots\ldots\ldots\ldots(9.2)$$

$$\ddot{V} = 8\pi R(\dot{R})^2 + 4\pi R^2(\ddot{R})\ldots\ldots\ldots\ldots\ldots\ldots\ldots\ldots\ldots\ldots\ldots(9.3)$$

For purposes of determining global expansion, the Hubble interior will be considered as minute volumetric divergences, each defined as fractional change in volume per unit area as area approaches zero. For an expanding empty void, the rate of expansion can thus be analogized to a dynamic modulus. To utilize Hubble values as representative of a larger universe, ratios between parameters determined within the Hubble sphere will be taken as indicative of the universe as a whole. All spheres of sufficient sample size would thus yield the same average density. Likewise, the fractional growth of Hubble volume in relation to its existing size should be constant for similar spheres. Thus, although the dynamic modulus is based upon a scalar density function derived from Hubble parameters, it is not limited thereto. The parameter itself is a ratio of factors each presumed equally applicable to an unlimited universe of any extent.

The divergence theorem relates the integral over the volume of the surface that contains the divergences to the flux exiting across the surface that contains the volume. By encompassing the expanding Hubble sphere with a Gaussian surround 'S' then from (9.3), since $R_S = R_3 = R_H$ at coincidence, we can drop the subscripts and write:

$$\frac{\ddot{V}}{Area} = \frac{8\pi R(\dot{R})^2 + 4\pi R^2(\ddot{R})}{4\pi R^2} = \frac{2(\dot{R})^2 + R(\ddot{R})^2}{R} \qquad (10)$$

15

The parameter devised to express changes in cosmological expansion velocity in terms of Hubble factors is given the symbolic representation 'q' defined as:[6]

$$q = -\frac{\ddot{R}R}{(\dot{R})^2} \qquad (11)$$

The 'q' factor is positive for a decelerating universe and negative for an accelerating universe. For an empty accelerating universe (*a la* de Sitter), or a zero energy universe where negative pressure energy equals positive $M_u c^2$ energy [as would be the case in (1) if $P = \rho_u c^2/3$], expansion is exponential and $q = -1$. For reasons later developed, the present state of the universe will be taken to be exponentially expanding thence $q = -1$. Upon substitution of (11) into (10):

$$\frac{\ddot{V}}{Area} = 3H^2 R \qquad (12)$$

For a static universe, Einstein's counter force $\Lambda R/3$ must equal the gravitational field at the limit of the Hubble sphere:

$$\frac{\Lambda R}{3} - \frac{GM_U}{R^2} = 0 \qquad (13)$$

Thus, for a Hubble mass M_u, the gravitational potential φ at the Hubble limit is:

$$\varphi = M_u G/R = 4\pi G(\rho_u)R^2/3 \qquad (13)$$

and therefore

$$4\pi G(\rho_u)R/3 = \Lambda R/3 \qquad (14)$$

From Friedmann's equation:[7]

$$4\pi G(\rho_u) = -q(3H^2) \qquad (15)$$

Hence from (14) and (15),

$$\Lambda = 3H^2 \qquad (16)$$

From (12), volumetric expansion per unit area directionally resolves along any axis as $H^2 R$, i.e., spatial expansion and Einstein's hypothesized cosmological

---

[6] After the discovery of the velocity distance law circa 1928, and throughout most of the 20th Century, expansion was assumed to be slowing due to gravitational retardation. To express the rate of change in terms of the Hubble parameters velocity and distance, a 'q' factor was concocted with a minus sign and given the name deceleration parameter.

[7] Appendix 10

force $\Lambda R/3$ are one-in-the-same. $\mathbf{G}$ and Λ cannot be interpreted as separate balanced forces, rather the present state of cosmological expansion determines the "now" value of $\mathbf{G}$. Expansion is the source and $\mathbf{G}$ is the consequence. That they are related derives from the reaction of the one upon the other. Einstein's 1916 addition of Λ to the General Theory brings about a unification only possible within a self creating cosmology. Expanding space is natures implementation of Einstein's cosmological constant; from Λ emerges $\mathbf{G}$ and from the action of $\mathbf{G}$ upon local matter, reactionary '$\mathbf{g}$' fields emerge. That is the subject to be next addressed. To restate, the relationship between volumetric expansion and Λ is:

$$[(d^2V)/dt^2]/4\pi R^2 = \Lambda R = 3H^2R \qquad (17)$$

Unidirectional acceleration $\mathbf{a_u}$ applied to a universe governed by Newton's 2nd law imposes reaction forces upon its inertial contents. The Force acting upon a uniform spherical mass $\mathbf{M_e}$ is:

$$\mathbf{F_u = M_e(a_u)} \qquad (18)$$

For (18) to be a pseudo force, the primary acceleration field must be isotropic and reactionary force concentrically convergent upon the inertial center of $\mathbf{M_e}$. That such a condition is created by spatial expansion, will be developed as per (37).

Figure 3 depicts an isolated spherically uniform inertial mass $\mathbf{M_E}$ concentric with its own Hubble sphere. To measure the exit flux, $\mathbf{M_E}$ is encompassed by concentric Gaussian sphere of radius equal to the distance where measurements are to be taken.[8]

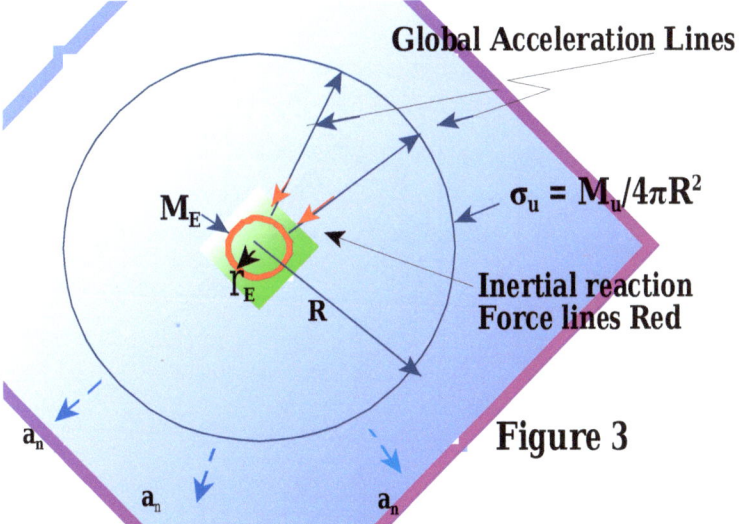

Figure 3

To illustrate let $\mathbf{M_e}$ and $\mathbf{r_e}$ be the mass and radius of a well-known object, specifically the green sphere is the earth and the red circle a gaussian surface

---

[8]To measure the reactionary field of the earth at its surface, the Gaussian surround would be coincident therewith. To measure earth's field at the moon, the Gaussian sphere would have a radius of 240K miles.

having the same radius r_E as the earth. From (18), Newton's 2nd law is converted to units of pressure and surface density by dividing the mass-acceleration product by its surface area A_E [which is also equal to the surface area of the gaussian surround]. Then for M_E entirely contained within the gaussian shell, the action of the expansion field at earth's surface is:

$$\frac{F_E}{A_E} = \frac{M_E}{A_E} a_N = \sigma_E a_N = P_E \qquad (19)$$

Where P_E is the negative pressure created by spatial expansion and a_n is the cosmological acceleration factor [volumetric acceleration per unit area from (12) resolved along a particular line of action]. The effect of the earth acting upon the cosmos is:

$$\frac{F_U}{A_U} = \frac{M_U}{A_U} g = \sigma_U g = P_N \qquad (20)$$

In (19), earth's volumetric density gets conveniently transformed to a shell density σ_E and in (20) Hubble volumetric density transforms as σ_U. These formulization(s) permit local reactionary fields to be specified in terms of global parameters. Specifically, local 'g' fields are the reactionary result of expansion created isotropic spatial acceleration whereas the action of expansion operating upon σ_U creates the global **G** field. The gravitational mechanics between an individual mass M_E and Hubble mass M_u is illustrated in **Figure 4**.

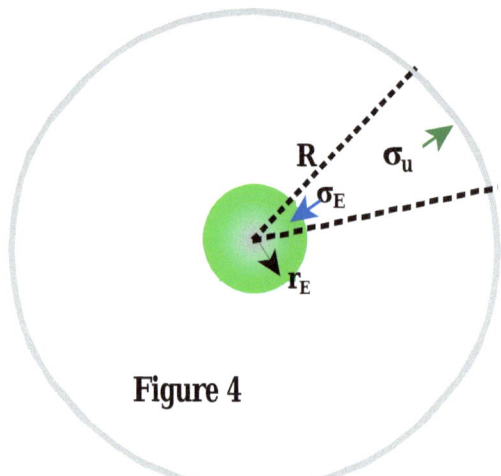

Figure 4

From Pascal's law, pressure is equal throughout space, so $P_E = P_n$, The intensity of the force per unit area at any distance 'r' greater than r_E is inversely proportional to the surface area $4\pi r^2$ of the imaginary gaussian sphere placed at distance 'r' over which the force is distributed. Since there are only two elements in our fictitious universe, then the reaction of the universe upon σ_E must be equal to the reaction of M_e upon the universe, and thus, for any distance 'r' from the center of M_E ($r_E < r < R$), reactionary counter acceleration will be proportional to the isotropic acceleration field divided by the area of the imaginary gaussian sphere $4\pi r^2$. From (19) and (20):

$$\frac{F_E}{A_E} = \frac{M_E}{A_E}a_N = \sigma_E a_N = P_E \quad = \quad \frac{F_U}{A_U} = \frac{M_U}{A_U}g = \sigma_U g = P_N \tag{21}$$

$$\frac{\mathbf{g}}{\mathbf{a_n}} = \frac{\sigma_E}{\sigma_U} = \frac{\dfrac{M_E}{4\pi(r_E)^2}}{\dfrac{M_u}{4\pi(R)^2}} = \frac{\dfrac{M_E}{4\pi(r_E)^2}}{\dfrac{kgm}{meters^2}} \tag{22}$$

Then
$$g = (\sigma_E/\sigma_U)a_N \ = \ \frac{a_N}{4\pi r^2}\left[\frac{M_E}{\sigma_U}\right] = \frac{c^2}{4\pi R\sigma_U}\ x\ \left[\frac{M_E}{r^2}\right] \tag{23}$$

Identifying the form of the equation as Newton's expression for the local Force, then

$$G = \frac{c^2}{4\pi R(\sigma_U)} \tag{24}$$

and (23) reduces to

$$g = G\frac{M_E}{r^2} \tag{25}$$

That (24) is a logical statement of Newton's constant, we dissect (24) into its parts for separate consideration per (27) below. At this point we comment upon the $\mathbf{c^2/R}$ term appropriated from (17). This corresponds to a $\mathbf{q}$ factor (-1) considered to be the present state of expansion. Exponential growth is the natural result of geometric doubling; a cubic unit of space [1 x 1 x 1] extends in each dimension creating [2 x 2 x 2] volume. While it is not known whether exponential growth has always been the order of the day, the curious reader may find Appendix A-8 of interest. While our objective is to relate natures long range fields to the present rate of expansion, some evolutionary history cannot be avoided. The "now" state of the universe is what it is because of what it was. To explain the present value of $\mathbf{G}$, requires a rationale for σ_U.[9]

---

[9]Following Newton's discovery of the laws of motion, the relationship between mass, force and acceleration was precisely measured by different types of experiments. In all cases, they proved to be insensitive to both the constitution and/or geometry of the masses. By contrast, gravitational attraction clearly depends upon distance. The laws seemed to be based upon different physics, yet the mass factor entered both formalisms with the same value. Einstein's resolution, "Equivalence" endowed mass with the power to modify space, but it did not address the magnitude or cause of the $\mathbf{G}$ factor.

When the 17th Century experimenters selected the unit of acceleration as the ratio of one unit of force to one unit of mass, they were dimensionally connecting space, time and mass. The measured values revealed an important essence of the universe as a whole. Whether cosmic mass-energy was acquired gradually or abruptly is not a factor in the dimensionality process; expressing 'G' in terms of global parameters is unrelated to the issue of whether cosmic mass-energy M_u was gradual or instantiated. However, in a self creating adiabatic cosmology, positive inertial mass would be required to increase in order to meet the volumetric growth of negative energy. Positive pressures lose energy during expansion. Since Negative pressure creates energy during expansion, it is expected that negative pressure remains constant in a zero energy universe[10]

Taking σ_U as "**1 kgm/m^2**" and separating the factors of (24) as:

$$G = \frac{c^2}{R} \times \frac{4\pi R^2}{M_U} \times \frac{1}{4\pi} \qquad (26)$$

The first term represents cosmological acceleration *a la* de Sitter, the second ratio is the **1/σ_U** factor (Hubble manifold area divided by Hubble mass), and the 3rd factor reflects the area over which the acceleration is spread when it acts upon another mass [from a logical perspective, the **1/4π** factor should be associated with '**g**' rather than '**G**' as it is arises from the area over which the reactionary force is spread]. Equation (26) thus reduces to:

$$G = \frac{c^2 R}{M_U} \qquad (27)$$

Newton's gravitational constant **G** reduces to Hubble volumetric acceleration divided by Hubble mass (**M_U** includes all forms of energy). **G** totes dimensions corresponding to *"volumetric acceleration per unit mass"* [**(meters^3/sec^2)/kgm**]. While **G** is composed of Hubble factors, it is not an exclusive property thereof. Any size sphere sufficiently large to include a representative sample of mass per unit volume will suffice. It should not be surprising that a global property of the universe is predicated upon global mass and volume. Why should it be otherwise?

In (27) it is understood **R** corresponds to the 2-sphere value **R_2** specified in (4) from which σ_U is devised. As a check, substitution of **1.45 x 10^{53} kgm** for **M_u** and **1.08 x 10^{26}** meters for **R**, then

$$\mathbf{G = [(3 \times 10^8)^2 \times (1.08 \times 10^{26})]/(1.45 \times 10^{53})}$$

$$\mathbf{= 6.7 \times 10^{-11} \ (m^3/sec^2)/kgm} \qquad (28)$$

---

[10] The first law of Thermodynamics [**dE + P(dV) = 0**] applies to both positive and negative pressure systems: When cosmic pressure is negative, positive energy increases during expansion.

That (24) follows directly from Friedmann's original 1922 paper can be seen from Appendix (A-11),

$$\rho_u = -q(3H^2)/4\pi G \tag{29}$$

For de Sitter expansion, $q = -1$, substituting $3\sigma_U/R$ for ρ_u from (8), then:

$$G = 3H^2/4\pi(\rho_u) = c^2/4\pi R\sigma_U \tag{30}$$

Which is the same as (24). A third approach that leads to the same result follows directly from the application of Newton's second law of motion to his law of gravitation. Consider a test mass M_x placed immediately beyond the Hubble manifold. By a bit of chicanery, one applies the velocity distance law to figure the acceleration at the Hubble limit. From the velocity-distance law $v = Hr$, so $dv/dt = H(dr/dt) + r(dH/dt)$. At the Hubble limit, $r = R$, $dr/dt = c$, and for H is constant, then:

$$dv/dt = H(dr/dt) = Hc = c^2/R \tag{31}$$

Setting the gravitational force equal to the reactionary force,

$$M_x(M_U)G/R^2 = M_x(c^2/R) \tag{32}$$

Since $(\rho_u)(4/3)\pi R^3) = M_U$,

$$G = 3H^2/4\pi\rho_u \tag{33}$$

Figure 5 illustrates the convergent reactionary flux (inwardly directed arrows) that defines the earth's gravitational gradient. For illustrative purposes, all other matter scattered throughout the Hubble volume is lumped into the shell and ignored. Spatial expansion creates a counter reaction proportion to the mass M_E divided by the area over which the reactance is distributed. In a curious similitude, space can be imaged as expanding where stress is negative and contracting where stress is positive; the force between two masses being the interaction of their respective stress created spatial contraction fields. By example, a non-expanding mass M_E subjected to an expansion engendered acceleration field a_n creates tension stress $M_E(a_n)/4\pi r_E^2$ at the σ_E surface. At any other distance $r > r_E$, the intensity is reduced to $M_E(a_n)/4\pi r^2$

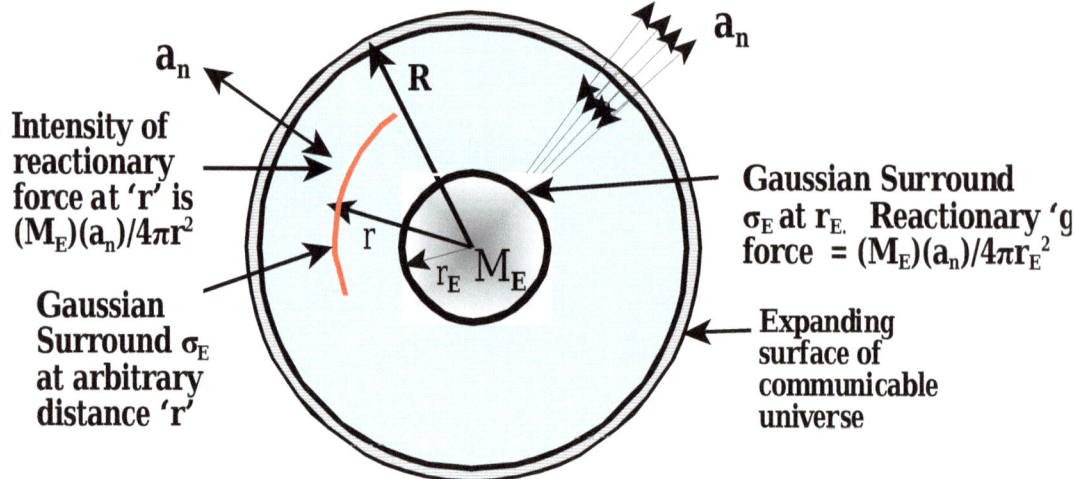

FIGURE 5

The First law of Thermodynamics (footnote 7) for negative pressure expressed as:

$$dE = -P(dV) = +(\sigma_u)(a_n)(dV) = +(\sigma_u)(a_n)4\pi r^2(dr) \qquad (35)$$

Whence, the Force:
$$dE/dr = +(\sigma_u)(a_n)4\pi r^2 \qquad (36)$$

At the Hubble limit, 'r' = 'R' and $\sigma_U = M_U/4\pi R^2$. The cosmological reactionary force F_R is therefore:

$$F_R = dE/dr = M_u a_n \qquad (37)$$

Taking (37) as Newton's 2nd law for an expanding universe, we pose the question of whether the Hubble sphere qualifies as a black hole. To study the universe from a Schwarzschild perspective without dealing with the complexities of General Relativity, the black hole equation is derived by equating the gravitational potential of the black hole mass **M*** to the kinetic energy at a radius r_s.[11]

$$(GmM^*)/r_s = v^2m/2 \qquad (38)$$

For light, (38) reduces to:
$$r_s = 2GM^*/c^2 \qquad (39)$$

---

[11] Shortly before he died as a solder in WWI, Karl Schwarzschild found the fist exact solution of General Relativity which covered the exterior geometry of a uniform spherical body. As with all equations devised from Special and General Theory, the black hole radius can be determined from the time-dilation formulation $\Delta t^* = \Delta t(1-2GM/rc^2)^{1/2}$ where **2GM/r** is the square of the escape velocity v_s. When the escape velocity is '**c**' or greater, equation (39) follows from (38).

For the universe, $r_s = R$. But what goes into the mass factor. All matter beyond the Hubble can be ignored, and all energy E within the Hubble in any form can be expressed as a mass equivalent $M_u = E/c^2$.

$$R_s = 2GM_u/c^2 \qquad (40)$$

Nothing is known about the interior field within a black hole, so the notion of singularity cannot be addressed. What does need to be considered is the affect of negative gravitational potential. Although representing 50% of Hubble mass, it appears to have no affect upon the radius R_s This reduces M_u by a factor of 2, whence:

$$GM_U/Rc^2 = 1 \qquad (41)$$

The ratio (41) was extensively studied by Carl Brans and Robert Dicke, in their search for a scalar-tensor theory of gravity.[12] Modeling the "now" state of the Hubble universe as an expanding black hole offers new insight re (41).

Interest in scalar-tensor theory was aroused following the publication of Paul Dirac's large-number hypothesis (LNH) in 1937.[13] Reasoning that the near equality of the *electro-gravitational* force ratio to the *cosmic-subatomic* size ratio was more than a coincidence, Dirac proposed these numbers had a cosmological connection. If true, the ratio of one to the other would be constant throughout the evolutionary history of the cosmos. As a consequence, the value of G would diminish inversely with the age of the Hubble universe. The general physics community discounted the LNH and other variable G theories based upon studies confirming the long term stability of lunar orbits. Unconsidered was the fact that orbital parameters depend upon the MG product, rather than G alone. If the orbiting mass m is small compared to the central body $M*$, then the approximate radius 'r' and orbit velocity 'v' follow from:

$$GM*m/r^2 = mv^{2/}r \qquad (42)$$

$$GM* \approx v^2r \qquad (43)$$

---

[12] The scalar–tensor gravity theory merges the curvature background of General Relativity with a scalar function that can vary in both space and time. Dicke and Brans used the concept to investigate the ratio (39), wherein the numerator was proposed to define inertial matter and the denominator its gravitational counterpart. The object was to show the ratio to be temporally invariant. As an adjunct of his large number hypothesis, Paul Dirac advanced the proposition that G varied as $1/R$, but observations of orbital motions were interpreted as evidence to the contrary. Several factors had to vary to simultaneously.

[13] Dirac occupied the Post of Lucasian Professor of Mathematics at Cambridge, a title once held by Isaac Newton and now by Stephen Hawking. Following his mathematical prediction of an anti particle having opposite polarity to that of the electron, the ratio of the electrical force between an electron and positron was established to be greater than the gravitational force of attraction by a factor of 10^{42}.

For orbital stability, the **GM*** product must be invariant. The problem of Dicke and Brans is also our problem [**R** as a factor in the denominator of equation (30)]. For long term orbital stability, a variance in either **G** or **M*** requires a compensating variance in the other. While cosmologists may be willing to accept a variation in one parameter, the defender's of the "Standard Model" are not inclined to embrace a theory that rattles the entire structure. Nonetheless, there is good reason for inertial mass and gravitational acceleration to be anti-variant, not the least of which is that it gives the universe "intelligible physical coherence."[14]

Gravity is detected by its influence upon the inertial content of a mass. Any mass having a present value **M** multiplied by the 'now' value of **G**, specifies the volumetric reactionary acceleration field of **M**. For the Hubble, from (33):

$$\mathbf{M_u G = \rho_u(V)G = c^2 R} \tag{44}$$

From (30): $$\mathbf{M_u = 4\pi R^2 \sigma_U} \tag{45}$$

Consistent with (44), the **M$_u$G** product increases proportionate to **R** whereas per (45), Hubble inertial mass increases as **R^2**. Inertia is perpetually generated by expansion, consequently there are no initial singularities and no breakdown of physical laws. Although **R** approaches zero when the time clock of the cosmos is wound backward, mass-energy also approaches zero. The idea of inertial mass as a "conserved quantity" has always been suspect. Gradual enhancement is a requirement of net zero energy. In what can be imprecisely referred to as "the beginning," the Hubble scale '**r**' was much less than its present **R**, consequently **c^2/r** expansion of the initial condition (**r << R**) creates the stress required for particle creation. As developed in Appendix (A-8), mass in the form of radiation and particles is a natural result of the intense stress created during the first jiffies of continuous **c^2/r** expansion. No inflationary interlude is required, nor is there need to tinker with the expansion profile, one continuous **c^2/r** function will suffice from beginning to present. Acceleration and stress diminishes smoothly as **1/r**. Subsequent recombination follows, the released photons (now observed as the CBR) being the signature of the abrupt beginning required by the "Standard Model." The stress field (now identified with **G**) continues to abate as inertial mass increases, all of which leads to the present state of affairs.

Expression (45), does not of itself, resolve the issue of whether expansion was past eternal or instantiated. As a curious aside, for a genesis universe the hot dense era can be emulated as a rapidly expanding electron. Specifically, for an initial scale **r$_0$** in the range of one Fermi [**10^{-15} meters**], the corresponding de Sitter acceleration state **c^2/r$_0$** will be **10^{42}** times greater than the present cosmological acceleration **c^2/R**. Spatial stress will be likewise greater than the present by a like factor. If inertial mass is proportional to surface area per (45), then the original electron fits the description of the Hubble sphere where size and mass have increases by **10^{42}**. Acceleration intensity diminishes by the same factor

---

[14] The phrase used by William McCrea to describe the **P = - ρ_uc^2** universe.

a la Dirac's LNH. The Hubble sphere, as Carl Sagan mused, may be but an expanding electron in a much larger universe.[15]

Figure 6 depicts the pressure intensity $a_n\sigma_E$ of the '**g**' field created by the dilating volume. The surfaces are projected as flat planes σ_U and σ_E [the density field $M_E/4\pi r^2$] viewed edge-on. Depending upon the viewpoint, the '**g**' field of M_E is a leftward acceleration [locations less than '**d**' where (c^2/r) equals gravitational counter reaction GM_E/r^2, virtual acceleration is toward M_E].

The intensity of the local '**g**' field associated with a uniform spherical mass **M** is thus related by three factors: 1) local mass **M**; 2) global inertia M_u; and 3) Hubble area $4\pi R^2$ Machian mass is thus an implicit component of local gravity (appearing as the ubiquitous scalar density field σ_U). Gravitational intensity increases linearly with distance until it reaches a maximum at the surface of M_E, then diminishes inversely with distance squared ($1/r^2$). In lieu of curvature, the volumetric acceleration field created by expansion. Whether considered as a dilating coordinate system or physically as isotropic spatial expansion, the dynamic theory of gravity constructs **G** as an intrinsic factor within the auspices of the expansion formalism.

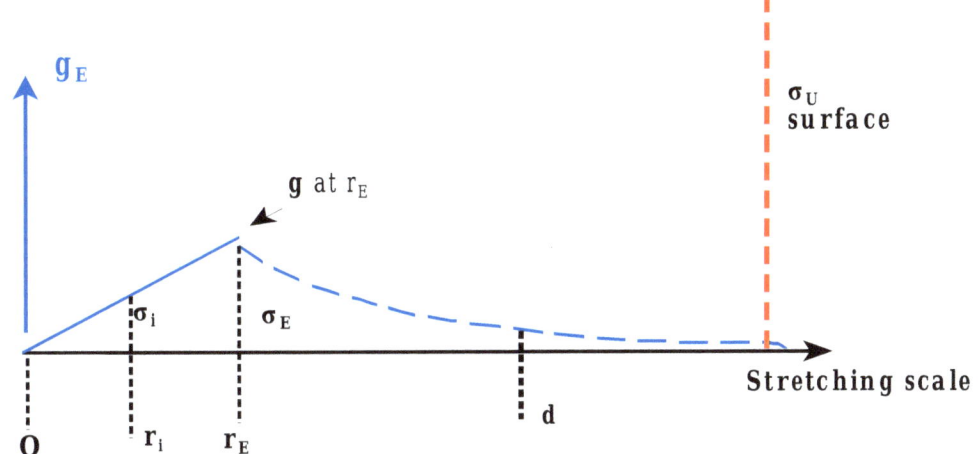

FIGURE 6

[15]Carl Sagan: "There is an idea...strange, haunting, evocative, one of the most exquisite conjectures in science or religion...an infinite hierarchy of universes, so that an elementary particle, such as an electron, in our universe, would, if penetrated, reveal itself to be an entire closed universe. Within it, organized into the local equivalent of galaxies and smaller structures, are an immense number of other, much tinier elementary particles, which are themselves universes at the next level, and so on forever---an infinite downward regression, universes within universes, endlessly. And upward as well. Our familiar universe of galaxies and stars, planets and people, would be a single elementary particle in the next universe up, the first step of another infinite regress."

To the extent the interior field intensity (solid blue line) is the result of expansion induced reaction, penetration of the spatial dilation function to the smallest subatomic scales is implied. The reactive field at the internal distance r_i is the result of all matter contained within r_i even as r_i is evaluated infinitesimally close to O. At the matter boundary r_E the slope intensity changes, but the strength of the field on both sides of the surface at r_E depends upon continuous expansion. The equations offer no distinction between gravity created by expansion of space within M_E and that created by expansion of space beyond. The matter boundary has significance only to extent of creating a discontinuity in the slope of the intensity profile (solid blue straight line to dotted blue curve). The expansion process would appear to be continuous at the interface between space and mass. [Metaphorically, it is as if matter is being stripped from the space it occupies]. Without a good description of space, however, all analogies fail as speculative. The importance is in the equation that describes the inertial reaction at the space-mass interface. Inertial reactance is instantaneous, it is reflected to the expanding source. Just as the σ_U scalar density field can be considered as located anywhere, the reflected reactionary field is manifest throughout the universe and encoded therein. All mass contained within the boundary defined by r_E is felt in the exterior space. From the perspective of the expanding coordinate system, local 'g' fields emerge as illusory spatial convergence. The gravitational binding limit 'd' for a mass M_E is obtained by equating the 'g' force of M_E to the cosmological acceleration c^2/R:

$$M_E G/r^2 = c^2/R \qquad (46)$$

Whence from (30) and (45):

$$d = r = R[M_E/M_U]^{1/2} \qquad (47)$$

While the intensity of the gravitational reactionary field diminishes inversely with distance squared, the totality of the field at any distance 'd' is spread over an area $4\pi d^2$, so total 'g' flux is the same at any distance 'd'. This unique property of natures long range forces, evidences continuous action. Reactionary 'g' fields emerge from the constant action of the primary G field acting upon expansion resistant matter.

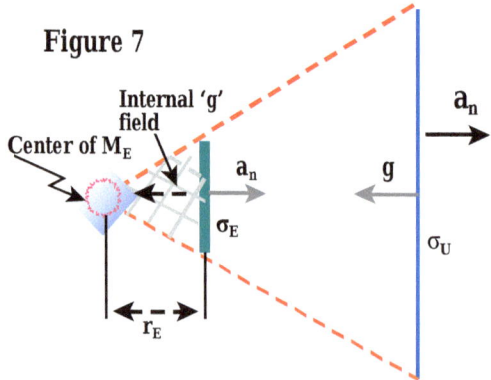

Figure 7

Figure 7 attempts to illustrate the action of the earth acting upon the universe in balance with the action of the universe acting upon the earth. This is the essence of (22). Expansion a_n acts upon Hubble mass [depicted as contained within the scalar density field σ_U] and earth [shown separated as the gray cross-hatched area]. The action of a_n upon earth's mass M_E creates its 'g' field. Beyond r_E, the intensity of the 'g' field diminishes inversely with the area over which it is spread as indicated by the dashed red lines extending to the Hubble limit where all other mass has been consolidated to make-up the scalar density σ_U. The g field created by the action of a_n upon the earth's mass M_E at r_E thus acts upon σ_U and by like reasoning, the mass of the universe M_u acts upon the earth with a pressure $a_n\sigma_U$. The surface density of the earth [$M_E/4\pi r_E^2$] is identified as σ_E. For an external gravitational field acting upon a spherically symmetrical mass, it is immaterial whether the mass is concentrated at a point or evenly spread over the surface of a 2-sphere, or distributed throughout the volume to create uniform density 3-sphere. Inertial reactance to an external gravity field depends only upon total mass. The inertial content of earth is the passive operative upon which the G field acts. [Whether some of earths inertia is due to gravitational interaction is immaterial].

Accordingly, the cosmological acceleration field acting upon the sigma surface density of the earth corresponds to the reactionary 'g' field created thereby, As previously, from (22)

$$a_n(\sigma_E) = g(\sigma_U) \tag{48}$$

Hence:

$$g_E = [M_E/4\pi r_E^2][c^2/R](\text{meters}^2/\text{kgm}) \tag{49}$$

For the earth, $M_E = 5.98 \times 10^{24}$ kgm and $r_E = 6.37 \times 10^6$ meters. Using the transformed value of $R = R_2 = 1.08 \times 10^{26}$ meters from (4), then (49) becomes:

$$g_E = [\sigma_E][c^2/R](\text{meters}^2/\text{kgm}) \tag{50}$$

Where $\sigma_E = 1.173 \times 10^{10}$ kgm/m^2, hence:

$$g_E = 9.8 \text{ meters/sec}^2 \tag{51}$$

Which is expected since the value of G derived in (24) is formularized from the same factors that make-up σ_E.

CLOSING THE GRAVITATIONAL LOOP

To complete the gravity scenario that began with Newton, requires a theoretic overhaul of what has evolved as the standard model. Indeed, the presumption of a universe initially endowed with mass-energy sufficient to fund all that is observed or yet to be manifest, is inconsistent with what is known of natural processes. To build a universe from nothing, requires time. Evolution must account for both the quantity of energy and size of space. To keep energy zero, expansion must augment negative and positive energy at the same rate. This should be taken as the 1st Principle of self-creation. Natural growth is a geometric process, the greater the quantity, the greater the rate at which more of the same is produced. Exponential volumetric expansion is consistent with what is known of physical processes.

The "standard model" has long been infected with a debilitating condition that has biased scientific outlook. It comes in two forms, 1) the once prevalent idea that the entire mass energy of the universe came into being as a singularity at the moment of its genesis, and 2) An inflationary phase of less than a fraction of a second where cosmic density remained constant during a period of rapid exponential expansion. Both were characterized as hot dense beginnings. That an early state of extreme stress existed at some juncture of cosmic history, is not in dispute. That such can exist in the form of newly created particles or enhanced inertia follows. An internally consistent zero energy universe, however, requires abandonment of the idea that mass comes "up front."[16] $\mathbf{G}$ depends from $\mathbf{c}^2/\mathbf{r}$ and $\mathbf{M}$ must increase proportionately as '$\mathbf{r}$' to maintain cosmic energy zero. The dimensional units of G betray it as a variable (volumetric acceleration should be expected to change as the volume changes). In actuality, de Sitter expansion leads to an increase in volumetric acceleration ($\mathbf{c}^2\mathbf{R}$) which would result in an increasing gravitational factor were it not for the fact that cosmic mass increases as radius squared ($4\pi\mathbf{R}^2$)[σ_U] per (45).[17] To restate, the inertial mass of an existing particle increases in proportion to $\mathbf{R}$, whereas total Hubble mass increases as the square of the radius per (45). From (24), $\mathbf{G}$ diminishes inversely with $\mathbf{R}$ consistent with (27) and (30).

---

[16] Edward P. Tryon, "Is the Universe a Vacuum Fluctuation?", *Nature*, vol. 246, p.396–397, 1973

[17] We might decide that there wasn't any singularity. The point is that the raw material doesn't really have to come from anywhere. When you have strong gravitational fields, they can create matter. It may be that there aren't really any quantities which are constant in time in the universe. The quantity of matter is not constant, because matter can be created or destroyed. But we might say that the energy of the universe would be constant, because when you create matter, you need to use energy. And in a sense the energy of the universe is constant; it is a constant whose value is zero. The positive energy of the matter is exactly balanced by the negative energy of the gravitational field. So the universe can start off with zero energy and still create matter. Obviously, the universe starts off at a certain time. Now you can ask: what sets the universe off. There doesn't really have to be any beginning to the universe. It might be that space and time together are like the surface of the Earth, but with two more dimensions, with degrees of latitude playing the role of time." -- Stephen Hawking, a Brief history of Time at page 129.

Considered next, a mass **M** on the surface of an empty Hubble sphere of radius **R** transported against the recessional acceleration of space c^2/R to the Hubble center. From Newton's 2nd law, the force **F** opposing the pilgrimage is:

$$\mathbf{F = Ma = M(c^2/R)}$$

Whence the energy is:

$$E = \int_{R}^{0} \frac{Mc^2}{R} dr = Mc^2 \tag{52}$$

Again, Newtonian physics upstages relativity in supplying a general expression for the energy content of a mass from Hubble values.[18] We now inquire into the rate of inertial accretion that follows from spatial expansion. Suppose the transport distance of **M** were **2R** instead of **R**? That is, for a Hubble sphere that has grown to twice its present size, the energy required for the trip is then **2Mc²**. In general, the effective energy associated with a particle is due to its relationship with the universe.

To understand gravity, it must be studied from a framework that does not include gravity.[19] When asked to describe General Relativity in one sentence, Einstein replied:

"Time, space and gravity have no separate existence from matter...physical objects are not in space, but these objects are spatially extended."

McCrea's matter creating algorithm is foundational to both *"steady state"* and *"genesis"* cosmologies. When the ($\rho_u c^2 = -P$) equation of state is plugged

---

[18] It is important to distinguish "gradually acquired Inertia" from the now discredited "Steady State Theory." The latter was originally proposed in 1939 by German Physicist, Pascal Jordan, having as its cornerstone the requirement of continuous particle creation. The idea was subsequently adopted by Fred Hoyle and other *"steady state"* theorists. For many, the idea of little-by-little creation offered aesthetic appeal. Moreover, it required no more credulity than what was claimed to have occurred at the instant of the big bang. *"Steady State Theory"* did not depend from a beginning and therefore lost favor after the discovery of CBR. By contrast, the theory of "gradually acquired inertia" embraces an era of intense stress created by c^2/r expansion where the expansion rate is inversely proportional to size. At the subatomic scale $r = r_0$ of the electron or less, particle creation follows as a logical consequence of the stress associated with rapid expansion of small volumes. The universe need not have a beginning, it can exist in a *"steady state"* of perpetual expansion that necessarily implies a smaller size in the past. In the last 13.8 billion years the universe would have grown from approximately one Fermi to its present value in the range of 10^{26} **meters**. The rate of expansion would have slowed from c^2/r_0 to c^2/R. Expansion of the universe thus comports with the diminution of the force field by a ratio of approximately 10^{41} which will be recognized as the difference in the strength of the **G** field compared to the strength of a one electron charge **q**.

[19] Standard theory begins with massless particles, at temperatures less than the electroweak scale (approximately 174 GeV) the universe is hypothesized to be in a symmetric stage. As the temperature drops, particles are theorized to acquire mass through the Higgs mechanism.

into the first law of thermodynamics, density during expansion remains constant, independent of volume, (a condition exploited by both Fred Hoyle and Alan Guth in bolstering their hypothesized theories). While McCrea's opus is well served by either scenario, both *steady state* and *inflation* contain omissions in critical areas where functionality is needed. Moreover, the generalized (**w = -1**) expansion state does not comport with zero cosmic energy. For inertial mass M_u equal negative gravitational potential, $\rho_u c^2$ equals **- 3P**. In a **q = -1** universe, $a_n = c^2/R$, then from (20),

$$3P = 3(\sigma_U a_n) = 3(\sigma_U)(c^2/R) \qquad (53)$$

And per (7), $\rho_u = 3\sigma_U/R$, hence:

$$\left| \rho_u c^2 \right| = \left| 3P \right| \qquad (54)$$

For de Sitter expansion, a pressurized Hubble volume has energy density $\rho_u(c^2)$. Coupled with the notion of continuous creation implicit in the disparaged *Steady State Theory,* inertial accretion finds a new ansatz in the form of increasing particle inertia following the short lived epoch in which they were created.[20]

Zero energy is the mandate self creation. Adiabatic expansion of negative pressure space corresponds to $\rho_u c^2 = -3P$ per (54). For a long past volumetric size comparable to that of an electron, the initial acceleration rate is c^2/r_0. In the cosmic time of one second, the scale of space increases by:

$$\Delta S = \tfrac{1}{2} at^2 = [c^2/2r_0][1/2] \approx [10^{17}/2(10^{-15})][1/2] \approx 10^{31} \qquad (55)$$

The corresponding volumetric change is 10^{93}. The conditions for development of proto particles follow from McCrea's analytics. To maintain cosmic balance, expansion must continue indefinitely. Internal stress diminishes from c^2/r_0 to c^2/R, (the denominator changing from 10^{-15} to 10^{26} meters. (Approximately the same factor predicted by Paul Dirac in his LNH.)[21] There are no discontinuities in the expansion state, as stress diminishes particle creation is replaced by inertial enhancement of existing particles. When volumetric acceleration has fallen below that necessary to create new particles, the inertia of existing particles continues to augment per (45), a necessary condition for zero energy. Perpetual creation of positive mass in the form of enhanced inertia proportionate to R^2 renders the illusion of fine-tuned initial conditions. The present Hubble density appears delicately balanced between run-away acceleration and gravitational collapse. The

---

[20] "Steady state theory" is consistent with the idea of continuous inertial enhancement following an initial epoch of sufficient acceleration to distill particles from spatial stress.

[21] If the universe originated from an electron, the field intensity would likewise have diminished by the same factor (10^{41}). The field would have only one polarity, as does gravity. A universe expanded from a positron would have opposite polarity. In both cases the force would be internally attractive due to the fact that the field is determined by expansion rather than contraction.

illusion of a fine tuned universe evaporates with the realization that the *theory of spatial expansion* and the *theory of inertial accretion* are Gemini twins, each dependent upon the other and inseparable there-from.

The vestiges of the original "Steady State Theory" are thus seen to be applicable in part to an eternally accelerating universe. As voiced by Fred Hoyle in 1975, *adjusting the rate of inertial accretion dissolves the usual mysteries as to how the universe begin.* Substitution of an electron as the state of earlier period, however, reclassifies the cosmology from *steady state* to genesis. Although the beginning need not necessarily be a Big Bang singularity, it is nonetheless a temporal beginning. We are then lead to questions as to whether an infinite initial acceleration for an infinitesimally short duration can be applied to zero point beginning, all of which brings to mind the intriguing scenario of an infinite number of universes each contained within a larger cosmic system as eloquently pictured by Carl Sagan in a previous footnote, and then the question of whether a zero sized universe requires an infinite time to grow from nothing to finite?

HARD TO BELIEVE PINCOCK, BUT FROM THE MATHEMATICAL PERSPECTIVE, THE UNIVERSE BOILS DOWN TO NOTHING GOING GOD KNOWS WHERE IN EVERY DIRECTION

THE INFLUENCE OF GRAVITY UPON TIME

In Einstein's world, *"gravity"* is curvature and *"time"* is relative. In General Relativity, the passage of *"time"* in a gravitational field depends upon gravitational potential. As between two static clocks at different heights, the high altitude clock runs faster. By contrast, Special Relativity treats all inertial frames equivalent, ergo, observers in relative uniform motion should, according to the tenants of Special Relativity, each measure time passing slower for the other guy. Such an experiment has never been conducted. Tests made to confirm time dilation take earth as the non-moving inertial frame, and in most experiments, the moving clock runs slower (as judged by clocks at rest on the earths surface). If a clock '**A**' has less velocity than a clock '**B**' at rest on the earths rotating surface the '**B**' clock will run slower.

Historically, the propagation of light and gravity were assumed to require some sort of ether to mediate transmission. Rene Descartes, John Bernoulli, Leonhard Euler, James Clerk Maxwell and numerous others, proposed a variety of structural models for the ether, all of which set the stage for the experimental endeavors that were later made to measure the earths speed relative thereto. The most significant of these were carried out by E.A, Moreley and Albert Michelson in 1887. While the set-up could only measure changes in the round trip velocity of light, the null results were taken as evidence of ethereal non-existence. Whether Einstein relied upon these experiments to formulate Special Relativity is not certain. What is known is that the equations in Einstein's 1905 paper *"On The Electrodynamics of Moving Bodies"* were identical to those previously derived by Danish physicist Hendric Antoon Lorentz. Einstein, however, imbued *"time"* with a startling mien unforeseen by all who had previously pondered the problem. To assert without experimental evidence, that observers in relatively moving frames experienced different rates of ageing, was both courageous and outrages. His premise, that all inertial frames are equivalent, although imposed at the outset, and made a rationale for the conclusions, stemmed from his personal conviction that the measured speed light is constant in all inertial frames. While the concept of *"time"* as a local experience, does not require such a sweeping imposition, the idea was enticingly provocative. In Part IV he makes the claim that it would lead to the following <u>peculiar consequence</u>: *"If at points **A** and **B**.... stationary clocks, viewed in the stationary system, are synchronous; and if the clock at **A** is moved with velocity **v** along the line **AB** to **B**, then on its arrival at **B**, the two clocks no longer synchronize, but the clock moved from **A** to **B** lags behind the other which has remained at **B** by (½)t(v²/c²) ...t being the time occupied in the journey from **A** to **B**. It is at once apparent that this result still holds good if the clock moves from **A** to **B** along any polygonal line, and also when the points **A** and **B** coincide. If we assume the result proved for a polygonal line is also true for a continuously curved line, we arrive at this result. If one of two synchronous clocks at point **A** is moved in a closed curve with constant velocity until it returns to **A**, the journey lasting t seconds, then by the clock which has remained at rest, the traveled clock on its arrival at **A** will be (1/2)tv²/c² seconds slow."*

The issue addressed herein, is whether the two theories of time dilation have a common denominator. If the passage of time can be predicated upon the local energy state, is it necessary or even proper to assert light speed as constant in all frames? In his *gedanken*, Einstein does not expect the stationary 'B' clock to experience change, yet, once the 'A' clock is accelerated to its cruising velocity, either clock can be considered as inertial. Straightforward application of the transforms, leads to the conclusion that each clock runs slow with respect to the other. In a situation where time dilation is real, clocks can be later compared; there are no paradoxical outcomes. Either one clock is ahead of the other or they are equal. To the extent the prophesized equivalence of all inertial frames leads to confusion, millions of words have been written. Concern here is with real time changes that result from potential or kinetic differences.[22]

Figure 8 depicts a clock scenario for a closed curve travel path between Pl Earth and Pluto. Initially, 'A' clock and 'B" clock are at rest on a high mountain top 'J' and properly synchronized. 'A' clock is then taken aboard a rocketship which is given a horizontal acceleration resulting in a low altitude circular polar orbit of radius r_e (green) and orbital velocity v_o. Measurements are taken to determine the rate of passage of time between 'A' and 'B.' After being brought to rest at **J**, the spaceship is attached to a tether anchored midway between Pluto and earth at **K** and given a horizontal boost to a velocity $v \gg v_e$ (where v_e is the escape velocity). Because of the tether 'A' is caused to travel a large circular path (red) making a Pluto pass-by then continuing to earth where measurements are taken to assess the time logged by clocks 'A' and 'B' when 'A' returns to 'J.'

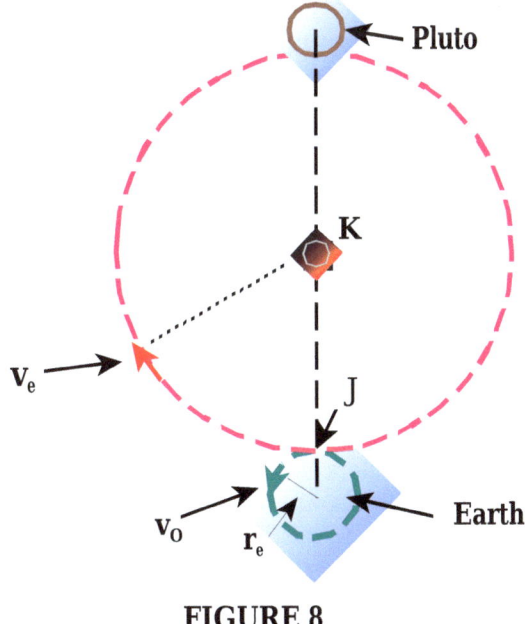

FIGURE 8

---

[22]Einstein's startling prediction that time depends upon velocity astonished the scientific world. The original (1905) description of a round trip experiment, would be later embroiled as "the clock paradox" Since either clock can be considered 'at-rest' then each observer can make the claim the other clock runs slow as judged by the passage of time in his own frame. energy.

It is known from numerous experiments, *"time"* for centrifuged clocks will be no different that that for a clock traveling at the same speed in a straight line. While acceleration can change velocity, acceleration *per se* does not affect *time dilation*. Centripetal acceleration does not alter velocity and therefore centrifuged clocks advance at the same rate as clocks in uniform linear motion.

In General Relativity, the dependence of *time* upon potential does not implicate curvature in the derivation. What is required is the global acceleration field '**G**.' Thus, in lieu of first relating gravity to mass induced curvature, pass can be made directly to a formulation of time in terms of energy. The fundamental hypothesis of the Special Theory (light speed constant in all inertial frames) deserves to be re-examined in the context of relative energies rather than relative velocities. The equivalence of Einstein's two expressions for *'relativistic time'* then follows from Minkowski's principle of *"Interval Invariance."*[23]

In the 1905 scenario, the traveling clock moves at constant velocity for the entire trip, the original scenario did not provoke the issue of time-lost at turn around There is no abrupt frame change. If earth's gravity is ignored, the time spent in following the dotted red path is Einstein's "peculiar consequence." Once the boost phase ends, cruising speed is constant – no energy is added or lost during the round trip excursion. Upon returning, **A** clock will be $(1/2)tv^2/c^2$ slow compared to **B** clock which has remained at **J**.[24] While **A** clock will have traveled in both *space* and *time*, **B** clock will have traveled only in *time* with respect to the earth frame with which they were both originally synchronized. From physics perspective, **A** clock has been accelerated to a higher kinetic state. The tether implements as a giant centrifuge. The time difference between **A** and **B** would, however, be the same if **A** had been placed in a laboratory centrifuge running at the same peripheral velocity '**v**' for the same amount of time '**t**.'

While the centrifuge can be considered non-inertial, and the difference between **A** and **B** clock readings vaguely rationalized as having something to do with General Relativity, this explains nothing. The rate at which **A** measures time is unaffected by the fact **A** travels a curved path. If **A** clock is not tethered – it leaves the earth in a straight line and continues. Tethering **A** facilitates does not affect the underlying cause of time dilation.

---

[23] Colloquially this can identified with the notion that *"all masses continuously move through spacetime with velocity '**c**' relative thereto."*

[24] Einstein's closed curve clock experiment aptly set the stage for Longevin's anthropomorphized version commonly known as the "twin paradox."[24] There is of course, no paradox. The rate at which clocks accumulate "time" and the rate at which biological systems age, both depend upon the same physics, namely the invariance of the spacetime interval measured in the cosmological rest frame. As it stands, one formalization for aging (Special Relativity) is based upon relative velocity and the other (General Relativity) depends upon absolute potential. In a 1918 paper, Einstein attempted to resolve the twin paradox by introducing a gravitational field to simulate turn around acceleration. Ironically, the original description involved a continuous curved path, so no turn around dilemma was involved. Nonetheless, many reputable physicists were led to later assert the "twin paradox" could not be analyzed without resort to General Relativity.

Bijective functionalities require correlative terms. The dependence of *"time"* upon velocity in Special Relativity and the dependence of *"time"* upon gravitational potential in General Relativity, suggests an underlying unity. The energies can be related provided the common reference frame is chosen to not offend either Theory. This is satisfied by a point at rest with respect to the CBR.

To make the curved path an inertial frame, a gravitational mass can be substituted for the tether [Falling trajectory and orbital motion are both valid inertial frames].[25] For this, the earth's mass $\mathbf{M_e}$ and radial size $\mathbf{r_e}$ could be taken as the central gravitational source and an orbiting mass $\mathbf{m_s} \ll \mathbf{M_e}$ will suffice as the orbiting spacecraft carrying clock 'A' to a Pluto flyby (ignoring all other mass in the solar system). The dotted green circular orbit about $\mathbf{M_e}$ is thus increased to encompass **Pluto**. Velocity $\mathbf{v}$ at radius R (not to be confused with Hubble radius $\mathbf{R}$), is then, per (43):

$$(\mathbf{M_e G})(\mathbf{m_s}/\mathbf{r}^2) = \mathbf{m_s}(\mathbf{v}^2/\mathbf{r}) \qquad (56)$$

And therefore
$$\mathbf{M_e G}/R = \mathbf{v}^2 \qquad (57)$$

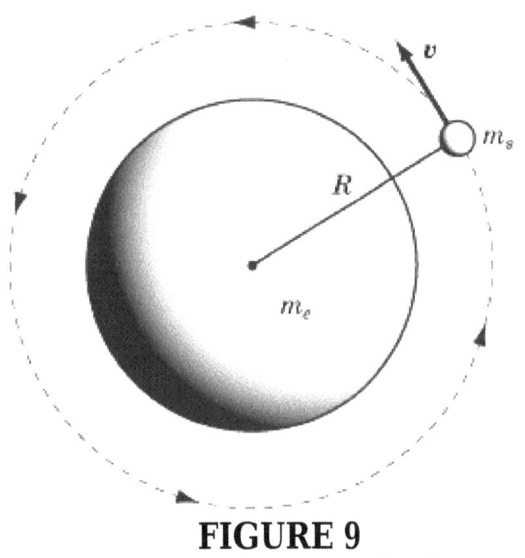

BETA $\rightarrow$

FIGURE 9

Both the local frame of $\mathbf{m_e}$ and Beta clock are considered not moving wrt the CBR. Beta clock is set to run at the same speed as a clock infinitely displaced from the local frame containing $\mathbf{m_e}$. $\mathbf{G}$ being a conservative field, total orbital energy (potential energy $\mathbf{E_p}$ plus kinetic energy $\mathbf{E_k}$) is constant, hence:[26]

$$- \mathbf{M_e G}/R + \mathbf{v}^2/2 = \mathbf{K_1} \qquad (58)$$

---

[25] The Road To Reality, by Roger Penrose, at page 394

[26] Equation (58) has several interesting extrapolations. In the Hubble limit, $\mathbf{r} = \mathbf{R}$, $\mathbf{v}^2/2$ is the square of the cosmic escape velocity [$\mathbf{c}^2$]. Hence (58) reduces to (41) if $\mathbf{M^*} = \mathbf{M_u}$:

$$\mathbf{G M_U}/\mathbf{Rc}^2 = 1$$

For any orbital distance, negative gravitational energy is greater than kinetic energy, so K_1 is negative.[27] When $v = v_e$, the mass m_s escapes the holding potential of M_e and $K_1 = 0$. There is no "time" slowing for $v = v_e$ since the two energies are equal and opposite. To make the circular path from earth to Pluto inertial, it is made gravitational. Consequently, "velocity-time-slowing" is nearly equal to "gravitational-time-slowing. The on-board satellite clock 'A' measures the passage of time in the inertial frame defined by the orbit whereas Beta clock measures cosmological time at infinity (The Hubble limit for present purposes). This time corresponds to the cosmological time for a non-moving clock at a point far removed from the influence of other matter. The orbital velocity v_0 is:

$$v_0 = (Gm_e/R)^{1/2} \tag{59}$$

The rate of passage of time due to the orbital motion of **A** relative to **B** is:

$$t_A = t_B[1-v_0^2/c^2]^{1/2} = t_B[1-GM_e/Rc^2] \tag{60}$$

The escape velocity v_e for a particle at distance R from the center of a uniform spherical mass m_e is

$$v_e = (2Gm_e/R)^{1/2} \tag{61}$$

And the rate of passage of time in the gravitational field of the earth's mass is:

$$T_A = t_B[1-2Gm_e/Rc^2]^{1/2} = t_B[1 - v_e^2/c^2] \tag{62}$$

The orbiting **A** clock experiences two time dilations relative to Beta clock. Either can be expressed in terms of kinetic or potential energy. The passage of time in any non-accelerating frame depends upon the local equation of state. Beta clock is taken as corresponding to a hypothetical zero energy state at infinity, ergo gravitational field energy due to M_e is negative. Time dilations are atoned, slowing depends upon energy, kinetic in the case of SR potential in the case GR.

Formulations (59) – (62) illustrate the correspondence between energy phenomenological(s). From (60), orbital kinetic energy $v_0^2/2$ can be expressed in terms the gravitational field and from (62) gravitational potential energy can be expressed in terms of kinetics (the velocity of escape v_e). The two formalisms are equivalent when both are referenced to the inertial frame of the universe as shown in **Figure 9**. The gravitational force acting upon m_s being opposite to the

---

[27] Here we have taken a point at the center of a mass as zero energy. Kinetic energy of motion is positive – if large enough it can be directed to eject m_s from the gravitational influence of the earth. In the situation posed, the kinetic energy of motion is sufficiently robust to separate the satellite mass from its earthly origin by a distance R, where it remains in orbit. To completely escape the earth's gravitational pull, more kinetic energy is required. The force pulling the satellite toward the earth is therefore opposite to the force produced by positive kinetic energy. Accordingly, the gravitational potential is negative.

centrifugal force produced by the kinetic energy associated with orbital velocity 'v,' net energy diminishes to zero when orbit velocity (v_o) equals (v_e).

Assume '**A**' has been accelerated until $v \gg v_e$ and now travels in a straight line at a uniform velocity relative to Beta. Clearly, there is no problem with '**A**' clock running slow wrt to Beta clock as judged by time on Beta clock. This is real time dilation that can verified when the clocks are brought together and compared (although comparison does not require clocks be brought together)[28] From the edicts of Special Relativity, Beta clock should run slow when measured by '**A**.' This, however, is inconsistent with the fact that '**A**' clock has already been verified as the loser. As between clocks, only one can be slow.

In actual practice, the fallacious idea that relatively moving observers each observe the other guy's clock to be running slow can be discredited by a simple experiment (previously described). After '**A**' and '**B**' clocks are synchronized in the laboratory, '**A**' clock is placed in a centrifuge which runs at constant peripheral velocity 'v' for a time 't.' Upon comparison '**A**' clock will have logged less time than '**B**' by $(1/2)tv^2/c^2$ (the same as if it traveled in a straight line or a continuously curved path as proposed by Einstein). The measured difference in logged time when the clocks are brought together and compared obviates any argument as to which clock ran slow. The reality of the measurement shows '**A**' ran slow during the time its energy state was elevated, a fact which cannot be turned around to fund the assertion that '**B**' ran slow from '**A**'s perspective.

While two spacecrafts moving in free space cannot determine their absolute velocities from their relative velocity, they can determine their energies based upon their launch velocities. If an earlier launched spacecraft is overtaken by a later launched spacecraft having greater velocity, the clock in the second spacecraft will run slower than the first as can be verified by signals sent back to earth. Any other determination of time dilation based upon relative velocity *a la* the transforms of Special Relativity, will be illusory.

The bottom line is that Beta clock always runs at the same rate. From an energy perspective, kinetic time dilation is on the same footing as gravitational time dilation, absolute and without ambiguity as to which clock runs slow. Energy differences are axiomatic, determinable with respect to a third frame from which both clocks have been launched. Einstein's great contribution was in the realization that *"time"* is local. To his holistic way of thinking, this meant symmetrical reciprocity. In reality, a more fundamental principle is in play, that which requires energy to be zero on the global scale.

---

[28]To avoid bringing the **A** and **B** clocks together to determine which clock aged most, a third clock '**D**' is initially placed at a distance '**L**' removed from earth. '**D**' clock is in the same energy frame as the earth (not moving with respect thereto) and therefore after placement, can be synchronized with '**B**'. The location of '**D**' is on the straight-line path followed by '**A**' after being launched from earth. When '**A**' clock passes '**D**' clock, it is read by an observer at '**D**' and compared to '**D**' clock which keeps the same time as '**B**' clock back on earth. The observer which holds the '**D**' clock will find the time logged by '**A**' to be less than that logged by '**D**' ergo, also less than that logged by '**B**' The clock which received the energy boost is unambiguously the clock that runs the slowest.

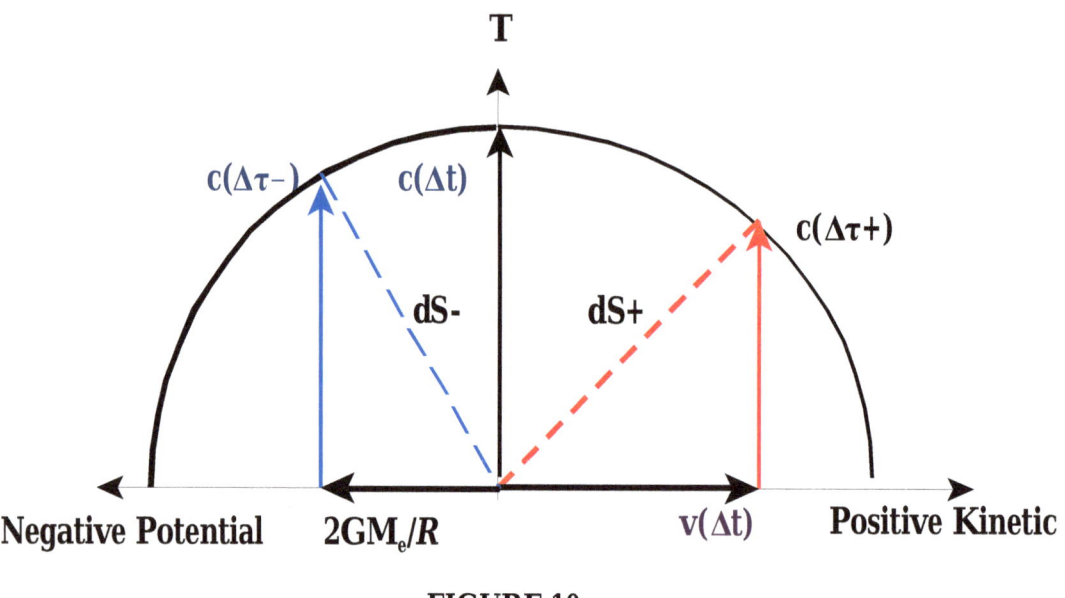

FIGURE 10

Figure 10. From the proposition that all objects move through spacetime at velocity 'c' then for an object moving with respect to the CBR at velocity 'v' the spatial distance is **v**(Δt) and the temporal distance is **c**($\Delta\tau$), whence the kinetic spacetime interval **dS+** is the *Pythagorean composite* cΔt = [(vΔt)^2 +[c$\Delta\tau$)^2]$^{1/2}$ Thus a local clock in free space moving at speed 'v' wrt the cosmological rest frame (CBR anisotropy = 0) will experience time as

$$\Delta\tau = \Delta t \ (1 - v^2/c^2)^{1/2}$$

Where 't' is cosmological rate of aging.

In a like manner, a clock suspended motionless in the negative potential of a gravitational field due to a mass **M$_e$** will run at a slower rate which corresponds to the virtual velocity **v$_e$** of space (**2GM$_e$/R**).   Net time dilation is the difference between time dilation caused by motion wrt space and time dilation caused by virtual motion of space wrt mass. For an object in free fall in a gravitational field, **v = v$_e$**, so time net dilation is zero.

Figure 11 depicts the 3-D expansion field metaphorically as an upwardly accelerating 2-D elastic fabric supporting an inertial mass **M**. Distortion of the fabric caused by the inertial reactance of **M** is the '**g**' field, the affect of the depression upon nearby matter indicated by the trajectories. *"Time"* slowing corresponds to the depth of the temporal depression whereas the '**g**' force at any radius '**d**' from the mass center is proportional to the spatial slope at '**d**.'

38

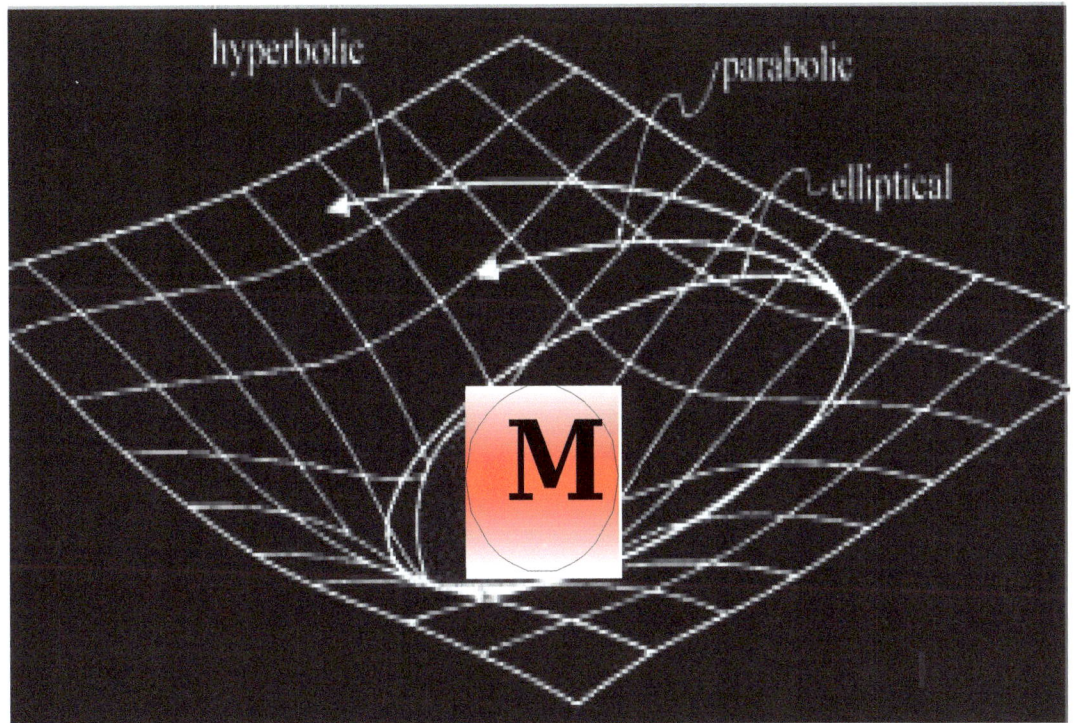

FIGURE 11

Epilogue

Gravitational forces are counter reactions that emerge from the reaction of local matter subjected to spatial expansion. Emergent theories of fields and forces are not new, but they have always been a dangerous proposition that provoked theologians. The preservation of his own life was likely the motivation for the guarded language and credit artfully bestowed upon the creator in this 17th century manuscript:

"...the action by which he (GOD) now sustains it is the same with that by which he originally created it; so that even although he had from the beginning given it no other form than chaos, provided only he had established certain laws of nature and had lent it his concurrence to enable it to act as it wont to do, it may be believed, without discredit to the miracle of creation, that in this way alone, things purely material might, in the course of time, have become such as we observe them at present; and their nature is much more easily conceived, when they are beheld coming in this manner gradually into existence, than when they are only considered as produced at once in a finished and perfect state."

Rene Descartes 1637

PISA, GALILEO'S BIRTHPLACE, AND ACCORDING TO THE APOCRYPHAL LEGEND, OBJECTS WERE DROPPED TO VERIFY ALL WEIGHTS FELL AT THE SAME RATE. THE 2002 PHOTO OF THE AUTHOR DISPELS A WIDELY HELD BELIEF THE TOWER IS LEANIING. IN QUESTIONS OF SCIENCE, BEING 'OFF-PLUMB' DEPENDS UPON THE PERSPECTIVE OF THE VIEWER

APPENDICES AND REFERENCES

Appendix (A-1):   Gravitational field C at a mass point produced by an infinite plane

The force on a unit mass at a given point **P** produced by a large sheet of material (Fig A-1), will be directed toward the sheet. Let the distance of the point from the sheet be '**a**' , and let the amount of mass per unit area of this large sheet be '**μ**' which is premised to be constant. What field dC is produced by the mass **dm** lying between **ρ** and **ρ+dρ** from the point **O** of the sheet nearest point **P**? Answer: $dC= -$ **G(dmr/r³)**. But this field is directed along r, and only the '**x**' component will remain when all the vectors dC are added to produce **C**. The **x** component of dC is:

$$\textbf{dCx} = - \textbf{Gdmrx/r}^3 \; = - \textbf{Gdma/r}^3$$

All masses **dm** which are at the same distance **r** from **P** will yield the same dC_x, so we may at once write for **dm** the total mass in the *ring* between **ρ** and **ρ+dρ**, namely **dm = μ2πρdρ** (**2πρdρ** is the area of a ring of radius **ρ** and width **dρ**, if **dρ ≪ρ**). Thus

$$\textbf{dC}_x = - \textbf{Gμ2πρdρa/r}^3$$

Then, since **r² = ρ² + a²**, **ρdρ = rdr**. Therefore,

$$\textbf{C}_x = -2\pi \textbf{Gμa} \int \textbf{dr/r}^2 \; = 2\pi \textbf{Gμa}(1/a - 1/\infty) = 2\pi \textbf{Gμ}.$$

Thus the force is independent of distance '**a**.' One might think that the farther away, the weaker the force.   But no! If **P** is close, most of the matter is pulling at an obtuse angle; if far away, more matter is situated favorably to exert a pull toward the plane. At any distance, the matter which is most effective lies in a certain cone. When farther away the force is smaller by the inverse square, but in the same cone, in the same angle, there is *more matter*, larger by the square of the distance.

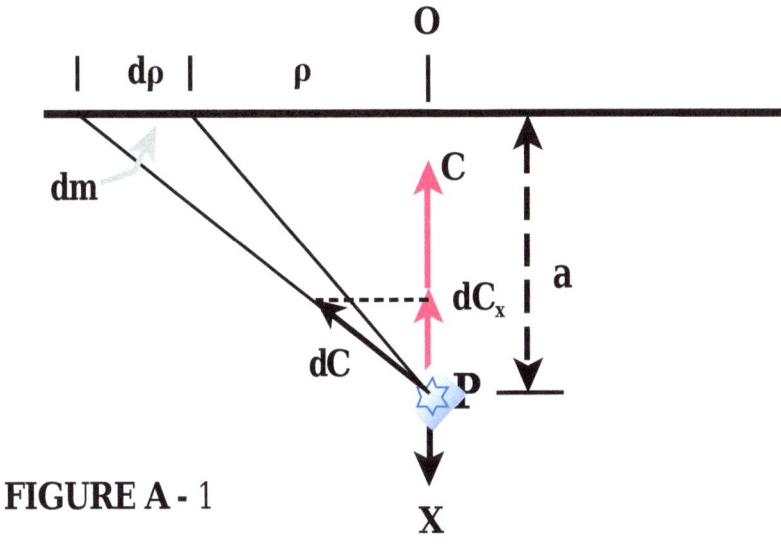

FIGURE A - 1

43

Appendix (A-2): Variable G.

The **R** term in the denominator of our expression (24) for **G** raises the issue of whether **G** is currently a variable. Long term studies of planetary lunar orbits are interpreted as proof of gravitational invariance. Equation (24)thus appears suspect. Like other variable **G** theories, the formulation presented here is consistent with Dirac's large number hypothesis. What is significant about the failed attempts to detect changes in **G** is that all such experiments only evidence the constancy of the mass-gravity product. Mass is a condition of energy. Energy is conserved, but mass converted to other forms of energy need not exhibit inertia, and when energy is in the form of mass, moving at relativistic velocity, greater inertia is observed. Mass is a state of energy, not a conserved quantity *per se*

All attempts to measure **G** are masked by the constancy of the **MG** product.[1] As our development predicts, an increase in space is accompanied by an increase in mass and consequently a decrease in the value of the gravitation coefficient **G,** but not a decrease in the magnitude of the negative pressure energy of the gravitational field. During non-accelerating expansion and constant **G**, negative pressure energy increases with volume R^3. But **G** decreases as **1/R, so** net negative Hubble energy increases as R^2. Thus, if σ_u, is the defining factor for inertia, negative energy increases at the same rate as Hubble manifold area $4\pi R^2$, so cosmic mass M_u can be written $4\pi R^2(\sigma_u)$. If the present state of expansion is exponential, **R** is constant, and the density of positive matter can be reckoned as equal to negative pressure energy in the gravitational fields of matter. Exponential de Sitter expansion freezes the Hubble scale and the Hubble distance **R** becomes a true horizon. There is thus compelling correlative evidence for a long term era whereby inertia increased as the cosmos expanded. Bottom line, the **R** in the denominator of (16) does not necessarily require **G** to be presently changing.[2, 3, 4]

---

[1]For an orbital moon, **v** and **r** depend upon the product of the planet's mass M_c multiplied by **G**. The orbital parameters **v** and **r** are determined by equating GM_cM^*/R^2 to centripetal force, M^*v^2/r from which:

$$GM_c = rc^2 \qquad (22)$$

[2]As an aside, gradually acquired inertia resolves a cosmological quandary, namely the high degree of tuning required for critical density. **G** has units of volumetric acceleration per unit mass [m^3/sec² **per kgm**]. To what might these units apply other than expanding space? And if applied to expanding space, is it not conceivable that the rate of volumetric growth would change as the size of the universe changes? The stability of orbits is testament to the invariance of the **MG** product. If **G** diminishes as **1/R**, the mass factor in the force equation must increase proportionately with **R**. Equation (2) can be viewed as other than a coincidental condition the present

[3]In 1937 Paul Dirac published the Large Number Hypothesis (**LNH**). Reasoning that the near equality between the electro/gravitational force ratio and Hubble/subatomic size ratio must be more than a coincidence, Dirac suggested that these large numbers maintain the same proportions at all times. This can only be true if one of the so called constants of nature changed as the universe expands. This lead to Dirac's hypothesis that **G** varies as ∝**(1/R)**.

[4]There is no law of conservation of mass. The inertial resistance of masses to acceleration increases for masses traveling at high velocities relative to the fame of measurement. Nor is their bases for the idea of an explosive mass creating genesis, although much effort has been directed to justifying such scenarios. Gradually acquired inertia is as it must be, it must grow to balance the negative energy of the expanding Hubble volume.

Appendix (A-3): The Faint Supernova Studies

Spontaneous creation has been a recurrent theme throughout scientific history. But an abrupt beginning of spatial expansion need not include the entire mass of the universe in a single event. The sudden appearance of mass-energy out of nothing is discrepant with all that is known about evolutionary process. Nonetheless, the general sentient of the twentieth Century had the more distant galaxies receiving a greater initial boost and therefore traveling farther since the beginning. The model was fortified by the belief recessional velocities were slowed by gravity, and for mainstream cosmology, exponential deceleration was the defacto standard for many years. The all at once matter myth requires expansion velocity to be fine tuned to avoid a quick crash or runaway expansion.

The 1998 supernova studies were based upon the proposition that SN bursts could be used as standard candles–the exclamation of identical energies, and therefore of equal brightness and duration. To the investigator's surprise, the intensity of the more distant events were fainter than what would be expected for a slowing universe. Either the universe was accelerating or something else was in play in the distant past.

The gravitational pressure needed to trigger a supernova was derived in 1932 by the Indian physicist, Subrahmanyan Chandrasekhar, for which he later received the Nobel prize.[5] The critical energy M_{limit} (approximately 1.4 solar masses) depends upon the factor $(hc/4\pi G)$. If **G** diminished inversely with **R**, the invariance of the **MG** product speaks directly to the question of whether supernova events were less energetic in the past. If that be so, the evidence for exponential expansion vanishes, and so also does the search for dark matter.[6] A larger **G** factor in the past requires less inertia to create the same force.[7] Since electron degeneracy pressure is constant, the inertial factor is less (Because **G** is greater in the past, less inertial matter is required to trigger a **1a** supernova event in the early universe). If intensity diminution is the result of less mass rather than greater distance, the theory of accelerating expansion needs to be re-thought. In this thesis, exponential cosmological expansion is the auspicate of declining **G** and its corollary, the gradual acquisition of inertia.

---

[5]A white dwarf star is kept stable by two opposing forces: 1) the electron degeneracy pressure created by nuclear fusion in the heart of the star (making lighter elements into heavier ones) pushing outwards from the core, and 2) gravity pulling inwards. When a white dwarf is locked in orbit with a companion star, it sucks off matter over time. This increases gravitational pressure until it overcomes the electron degeneracy pressure. The amount of mass in the core has a special significance called the Chandrasekhar Limit. When the core acquires a mass of approximately 1.4 solar masses, electron degeneracy pressure is overcome by the pressure of gravity.

[6]As a side note, efforts to explain the present value of **G** in terms of **q** = ½ led to much frustration for the author. The discovery of Cosmological Acceleration provided a good fit to the empirical value of **G** based upon standard model consensus H_0 = 71. The perception of uniformly expanding **3-D** space as 'time' can be appreciated as perspective on **4**th dimensionality.

[7]Because the **MG** product is constant, the weight of the mass required to overcome the degeneracy pressure is the same at all eras. Since the electron degeneracy pressure does not change with time, the **MG** pressure required to trigger a supernova event will also be invariant. A larger **G** during an earlier era translates to smaller **M**, and consequently less energetic events. Faint supernova in the distant past may have more than one explanation.

Appendix (A-4): The Faint Sun Syndrome

Geophysical and climatological data show the earths temperature during the past four billion years has not changed appreciably. However, numerical models based upon the Sun's interior indicate the Solar output would have been approximately 25% less than its present value. Various theories have arisen to justify the warm conditions that prevailed for the young earth. Of significance for this treatise, is the variable **G** theory.

The Sun's luminosity $L_\square$ is highly sensitive to **G** and **M**, being roughly proportional to $\mathbf{G^7M^5}$. The invariance of the **MG** product thus provides a mechanism and explanation for a temperate beginning. A 25% reduction in solar output based upon the Sun's condition and status as a main sequence star would be roughly balanced by a robust **G** and a smaller inertial mass **M**. In a recent publication, the variable **G** theory was studied and compared to alternatives founded upon atmospheric changes, most notably suppositions based upon unsupported levels of green house gases in the atmosphere of the young earth.[8] The authors avoided consideration of changes of in the range of **1/R** as proposed by Dirac' in his Large Number hypothesis ($\mathbf{G \propto t^{-1}}$). Consequently, they were able to explain their conclusions in terms of small changes. However, as developed herein, decreasing **G** is accompanied by increase in the inertial property of existing particles. The effect upon luminosity $L_\circ$ due to changes in **G** is reduced from 10^7 to 10^2 when the 10^5 effect of increased mass is factored into the equation.

---

[8]Can A Variable Gravitational Constant Resolve The Faint Young Sun Paradox?; International Journal of Modern Physics D. Varun Sahini, Yuri Shtanov, Nov 2014.

Appendix (A-5): Relativistic Conformance.

    Both the Special and General Theory of Relativity are integrally tied to the space/time ratio 'c'. Although rarely expressed in terms of energy transformations, real time dilations always involve reference frames with energy differences. For special Relativity, the energy difference is kinetic, the relative rate of time depending upon the kinetic energy $v^2/2$ where v is the relative velocity. In what initially appears to be a contrast between the Special and General Theory, the latter relates time dilation to gravitational potential. But upon closer examination, it will be understood that gravitational time dilation also depends from kinetic energy differences, specifically, the velocity required to escape the gravitational well:

$$\Delta t^* = \Delta t (1 - 2GM/rc^2)^{1/2} \tag{40}$$

where $2GM/rc^2$ is the escape velocity needed overcome the gravitational field of M, or viewed alternatively, the velocity acquired by an object falling from ∞ to the surface of a uniform spherical mass M of radius r. That this factor has a familiar complexion, suggests probative implications if applied to our development of G based upon M_u and R, that is, from (8) and (34):

$$\Delta t^* = \Delta t \left[1 - 2 \frac{4\pi R^2 \sigma_u c^2}{4\pi R \sigma_u (Rc^2)} \right]^{1/2} = \Delta t [-1]^{1/2} = i\Delta t \tag{41}$$

Consistent with uniform cosmological time, (41) reduces to one unit of temporal distance defined along the "i" axis $\perp$ to the 'x', 'y', and 'z' dimensions of 3-D space.

Appendix (A-6): Reconstructing Newton's Second Law

In the prelude to gravity, Newton's second law was adapted to the gravitational field by diluting the force over the active surface area (For the isotropic global acceleration field c^2/R, the appropriate area is the Hubble sphere $4\pi R^2$ and the appropriate mass is M_u. A local mass M_o subjected to acceleration experiences the reactance of the scalar density field σ_u Transformation from mass to unit area proceeds by dividing both sides of Newton's 2nd law by one square meter:

$$F/m^2 = (M_o/m^2)a$$

From $kgm/m^2 = \sigma_u$ then

$$F/kgm = (M_o/m^2)(a/\sigma_u)$$

Appendix (A-7) From Newton's 2nd Law—No Dark Energy Required

Newton's 2nd law for a matter creating expanding space is:

$$\mathbf{F = (d/dt)(Mv) = M(dv/dt) + v(dM/dt)}$$

wherefore since:

$$\mathbf{M_u = \rho_u(4/3)\pi R^3} \text{ then } \mathbf{dM_u/dt = \rho_u(4\pi R^2)(dR/dt)}$$

From which

$$\mathbf{F = (\rho_u)(4/3)(\pi R^3)(dv/dt) + (v)\rho_u(4\pi R^2)(dR/dt)}$$

Since $\mathbf{v = c}$ at the Hubble limit $\mathbf{R}$, then $\mathbf{dR/dt = c}$, and therefore if F = 0 on the global scale and density is constant during expansion, the Hoyle solution is

$$\frac{\mathbf{dv}}{\mathbf{dt}} = -\frac{\mathbf{v}^2(4\pi\mathbf{R}^2)\rho_u}{\dfrac{4}{3}(\pi\mathbf{R}^3)\rho_u} = -\frac{3c^2}{R}$$

which is consistent with $(\rho_u)c^2 = -\mathbf{P}$. This however is not a zero energy universe. In the exponentially accelerating zero energy universe, $(\rho_u)c^2 = (-3P) = -(3\sigma_u)(a_n)$, so:

$$\mathbf{a_n = [(\rho_u)c^2]/3\sigma_u = c^2/R}$$

Appendix (A-8): Perpetual Acceleration

Hubble radius is approximated by assuming expansion rate to be '**c**' on average over the life of the expansion process. Expressed mathematically

$$dS = c \, dt$$

The Hubble scale **R** is approximated as:

$$R = \int dS = \int_0^\tau c(dt)$$

Where 'τ' is the Hubble time (approx 13.8 billion years). This is the profile of a non-accelerating universe expanding from beginning to the present at constant velocity '**c**.'

Surprising, the same result can be obtained for the perpetually accelerating universe, provided time is taken as **1/H** and **H** is taken as **c/r** where '**r**' is the instantaneous value of the Hubble scale at time '**t**' Integration from '**0**' to 'τ' presents a problem for **c²/r** expansion at the lower limit where r → 0. However, substitution of **r/c** for the temporal variable leads to curious possibilities. The integration encounters no problems at the beginning where r → 0 because the expression for time '**r/c**' and the expression for acceleration **c²/r** are canceling. Specifically, for an acceleration '**a**' the incremental expansion distance **dS** during an increment of time **dt** is:

$$dS = at(dt)$$

For de Sitter expansion, **a** = **c²/r**, and since **t** = **1/H** = **r/c**, the distance '**S**' for a Hubble time interval 'τ' is:

$$S = \int dS = \int_0^\tau at(dt) = \int_0^\tau \frac{c^2}{r}\left[\frac{r}{c}\right]dt = \int_0^\tau c(dt) = R$$

The result is the same as that for a constant expansion rate '**c**.'

Because '*time*' was formulated as the inverse of the Hubble constant, the **r** component in the expression for the acceleration cancels with the **r** component in the expression for cosmic time. This leads to a linear dependence between Hubble size and Cosmic time which conforms closely to the empirical data. Because early acceleration is large and of short duration, proto particle formation is likely because stress during the first jiffies of expansion is maximum

Finally, there is a bazaar interpretation of above which suggests time to be a variable **r/c**. The smaller the radius the slower the passage of '*time*.' On the scale of expanding 'time,' the universe is past eternal, yet it displays the countenance of a genesis.

Appendix (A-9): Building the Universe from Scratch

Null Universe The kinetic energy **dE** of a shell of mass density ρ, area **A** and thickness **dr** is:

$$dE = (A\ dr)(\rho)(v^2/2)$$

and since **v=Hr**, and **A = 4πr²** the total kinetic energy **E** is:

$$=\frac{\rho}{2}\int_0^R 4\pi r^2 H^2 r^2 (dr)$$

$$=\frac{\rho H^2}{2}\left[\frac{4\pi R^5}{5}\right]$$

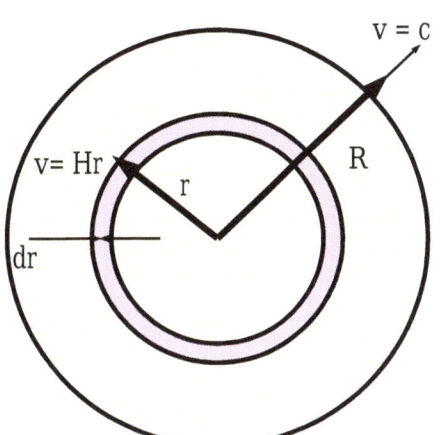

Figure 7a

If the Universe is assembled by building it from thin spherical layers of thickness **dr** as shown in **Fig 7b**, the differential work at each stage is:

$$dU = Gm_r(dM)/r$$

since **$M_r = \rho(4/3)(\pi r^3)$** then **$dM = \rho(4\pi r^2)dr$**; the work in bringing-up the universe is:

$$U \;=\; G\int_r^R \frac{16}{3}\pi^2(\rho_u)^2 r^4\,(dr)$$

$$=\frac{3G(M_u)^2}{5R}$$

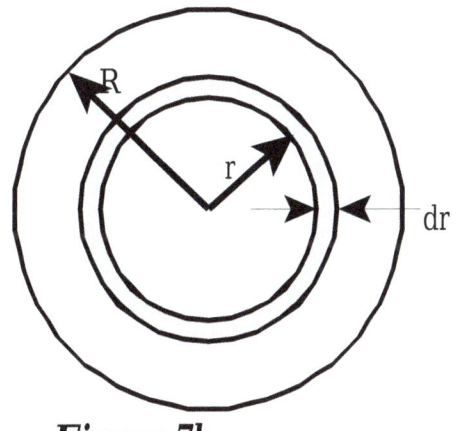

Figure 7b

In a null universe, positive kinetic energy balances negative potential:

$$\frac{\rho_u 4\pi H^2 R^5}{10} = \frac{3(\rho_u V)^2 G}{5R} \quad \therefore \quad \rho_u = \frac{3H^2}{8\pi G}$$

When the energy density is taken only as kinetic (**$mv^2/2$**) as in the first equation, the exponentially decelerating **q = ½** universe emerges. If energy had been taken instead as equal to the field energy (- **mc^2**) rather **$mv^2/2$** the corresponding density is doubled. This comports with de Sitter's accelerating universe **$(3H^2)/4\pi G$**.

Appendix (A-11): Friedmann Equations from Newtonian Physics

In its simplest form, the development starts with a uniform density sphere of fixed radius "**a**" where the escape speed is $v_e = 2GM/a$. The strength of the **G** field outside the sphere depends upon **M** and the distance "**r**" from its center, but not the radius "**a**." If **M** itself is considered as expanding so the surface particles have radial velocities, the expansion profile will follow the same trajectory as a particle launched normal to the surface. When the universe is modeled with a big bang beginning, the escape velocity analogy applies. Specifically, if the expansion velocity at the surface is less than the escape velocity, the sphere will slow and eventually begin to contract. If the launch velocity equals or exceeds v_e, expansion is eternal. The total energy is the sum of the kinetic and potential energy. If the sphere represents the universe, then at time t_0 the radius of the sphere will be "a_0" and the cosmic scale factor will be R_0. Therefore $a/a_0 = (R/R_0)$ so the cosmic radial velocity is $(a_0)[(dR/dt)/R_0]$, then:

$$\dot{R}^2 = \frac{8\pi G\rho R^2}{3} + C\left[\frac{R}{a}\right]^2$$

For a uniform density sphere, the second term is constant; it corresponds to the imagined orbits of ejected surface particles which can be elliptical, parabolic or hyperbolic as shown in **Figure 8a** of Chapter III. As applied to a universe where gravity is not an expansion dependent force, it predicts the ultimate cosmological fate (collapse or eternal expansion). For convenience, **R** can be scaled by a normalizing factor R_0 so that **CR/a** is represented by a single constant **(-k)** which takes the value **[+1]** for elliptical paths, **[0]** for parabolic orbits and **[-1]** for hyperbolic trajectories (eternal expansion of negatively curved space). In the Newtonian analogy **k** identifies the flight of an ejected surface particle whereas in General Relativity, **k** is the curvature constant, expressed in terms of the distance scale as $K = k/R^2$. The Hubble term $H = (dR/dt)/R$ and $q = -d^2R/RH^2$ Substitution in the above gives:

$$K = H^2(2q-1)$$

$$4\pi G\rho = 3qH^2$$

For **q = -1**,

$$G = -3H^2/4\pi(\rho_U)$$

www.ingramcontent.com/pod-product-compliance
Lightning Source LLC
Chambersburg PA
CBHW040744200526
45159CB00023B/1683